Electric Motors and Drives

Fundamentals, types and applications

Austin Hughes

Senior Lecturer, Department of Electrical and Electronic Engineering, University of Leeds

Heinemann Newnes

Heinemann Newnes
An imprint of Heninemann Professional Publishing Ltd
Halley Court, Jordan Hill, Oxford OX2 8EJ

OXFORD LONDON MELBOURNE AUCKLAND SINGAPORE
IBADAN NAIROBI GABORONE KINGSTON

First published 1990
Reprinted 1990

British Library Cataloguing in Publication Data
Hughes, Austin
    Electric motors and drives.
    1. Electric motors
    I. Title
    621.46'2

ISBN 0 434 90795 2

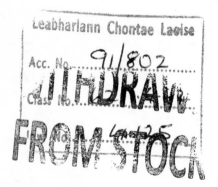
Typeset by BP Datagraphics Ltd, Bath, Avon
Printed in Great Britain by Billings of Worcester Ltd

CONTENTS

PREFACE

This book is intended for non-specialist users of electric motors and drives. It fills the gap between specialist text-books (which are pitched at a level which is too academic for the average user) and the more prosaic 'handbooks' which are filled with useful detail but provide little opportunity for the development of any real insight or understanding.

Exciting and radical changes have taken place in motors and drives over the past twenty years, and the contemporary scene is now very different from the old order. For more than a century, different types of motor continued to be developed, and each became closely associated with a particular application. Traction, for example, was seen as the exclusive preserve of the series d.c. motor, whereas the shunt d.c. motor, though outwardly indistinguishable, was seen as being quite unsuited to traction applications. The cage induction motor was (and still is) the most numerous type but was judged as being suited only to applications which called for constant speeds.

The reason for the plethora of motor types was that there was no easy way of varying the supply voltage and/or frequency to obtain speed control, and designers were therefore forced to seek ways of providing for control of speed within the motor itself. All sorts of ingenious arrangements and interconnections of the various motor windings were invented, but even the best motors had a limited range of operating characteristics, and all of them required bulky control gear which was manually or electromechanically operated, making it difficult to arrange automatic or remote control.

All this began to change from the early 1960s, when power electronics started to make a real impression. The first breakthrough came with the thyristor, which for the first time allowed an efficient, compact, and easily controlled variable-voltage supply to be employed in the main (armature) circuit of the d.c. motor. Some minor changes were called for in the motor to accommodate the less than perfect 'd.c.' which the controlled thyristor rectifier provided, but the flexibility offered by the electronic control of speed and torque meant that a truly adaptable workhorse was at last a practicable proposition. In the 1970s the second major breakthrough resulted from the development of variable-frequency inverters, capable of providing a three-phase supply suitable for induction motors. The induction motor, which for so long had been relegated to constant speed applications, was suddenly able to compete in the controlled-speed stakes.

These major developments resulted in the demise of many of the special motors, leaving the majority of applications in the hands of comparatively few types. In effect the emphasis has now shifted from complexity inside the motor to sophistication in supply and control arrangements. From the user's point of view this is a mixed blessing. He can look forward to greater flexibility and superior levels of performance, and there are fewer motor types for him to consider. But if anything more than constant speed is called for, the user will be faced with the purchase of a complete drive system, consisting of a motor together with its associated power electronics package. To choose wisely, he needs not only to know something about motors, but also to be able to assess the suitability of the associated power electronics, and to understand the control options which are normally provided.

An awareness of the user's potential difficulties has been

uppermost in shaping the contents of this book. The aim throughout has been to provide the reader with an understanding of how each motor and drive system works, in the belief that it is only by knowing what should happen that informed judgements and sound comparisons can be made. Given that the book is aimed at non-specialists from a range of disciplines, introductory material on motors and power electronics is clearly necessary, and this is presented in the first two chapters. Many of these basic ideas crop up frequently throughout the book, so unless the reader is already well-versed in the fundamentals it would be wise to absorb the first two chapters before tackling the later material.

The remainder of the book explores most of the widely-used modern types of motor and drive, including conventional and brushless d.c., induction motors (mains and inverter-fed), stepping motors, synchronous motors (mains and converter-fed) and reluctance motors. The d.c. motor drive and the induction motor drive are given most weight, reflecting their dominant position in terms of numbers. Understanding the d.c. drive is particularly important because it is so widely used as the yardstick by which other drives are measured. Users who develop a good grasp of the d.c. drive will find their know-how invaluable in dealing with all other types, particularly if they can establish a firm grip on the philosophy of the control scheme.

Applications are deliberately spread throughout the text in order to emphasize the fact that there is no longer any automatic correlation between motor type and particular application. Similarities between the various motors and drives have also been given emphasis in order to underline the fact that apparently different types have a great deal in common at the fundamental level. Recognizing this degree of commonality is important from the user's viewpoint, but is seldom given weight (albeit for obvious reasons) by enthusiastic salespersons.

Deciding on the style of the book was relatively easy, and it reflects my own preference for an informal approach, in which the difficulty of coming to grips with new ideas is not disguised. On the other hand, deciding on the level at which to pitch the material was a more difficult matter, given the broad range of backgrounds of potential readers. Experience suggested that a mainly descriptive approach with physical explanations would be most appropriate, so this pattern has been followed throughout, the space and depth of treatment being varied according to the importance of the topics. Mathematics has been kept to a minimum in the belief that this will make the contents easier to digest.

Austin Hughes

1

ELECTRIC MOTORS

INTRODUCTION

Electric motors are so much a part of everyday life that we seldom give them a second thought. When we switch on an electric drill, for example, we expect it to run rapidly up to the correct speed, and we don't question how it knows what speed to run at, nor how it is that once enough energy has been drawn from the supply to bring it up to speed, the power drawn falls to a very low level. When we put the drill to work it draws more power, and when we finish the power drawn from the mains reduces automatically, without intervention on our part.

The humble motor, consisting of nothing more than an arrangement of copper coils and steel laminations, is clearly rather a clever energy converter, which warrants serious consideration. By gaining a basic understanding of how the motor works, we will be able to appreciate its potential and its limitations, and (in later chapters) see how its already remarkable performance can be even further improved by the addition of external controls.

This chapter deals with the basic mechanisms of motor operation, so readers who are already familiar with such matters as magnetic flux, magnetic and electric circuits, torque, and motional e.m.f. can probably afford to skip most of it. In the course of the discussion, however, several very important general principles and guidelines emerge. These apply to all types of motor and are summarized in the concluding section. Experience shows that anyone who has a good grasp of these basic principles will be well equipped to weigh the pros and cons of the different types of motor, so all readers are urged to absorb them before tackling other parts of the book.

PRODUCING ROTATION

Nearly all motors exploit the force which is exerted on a current-carrying conductor placed in a magnetic field. The force can be demonstrated by placing a bar magnet near a wire carrying current, but anyone who tries the experiment will probably be disappointed to discover how feeble the force is, and will doubtless be left wondering how such an unpromising effect can be used to make effective motors.

We shall see that in order to make the most of the mechanism, we need to arrange for there to be a very strong magnetic field, and for it to interact with many conductors, each carrying as much current as possible. In some motors, it will be easy to differentiate between the parts of the motor which are responsible for setting up the magnetic field, and the parts carrying the conductors on which the forces act. In others, such as the induction motor, the physical distinction is not obvious, but we will find it helpful to think along the same lines to assist in understanding how such motors work.

We will look first at what determines the magnitude and direction of the force, before turning to ways in which the mechanism is exploited to produce rotation. The concept of the magnetic circuit will have to be explored, since this is central to understanding why motors have the shapes they do. Magnetic flux and flux density will crop up continuously in the discussion, so a brief introduction to the terms is included for those who are not already familiar with the ideas involved.

Electromagnetic force

When a current-carrying conductor is placed in a magnetic field, it experiences a force. Experiment shows that the magnitude of the force depends directly on the current in the wire, and the strength of the magnetic field, and that the force is greatest when the magnetic field is perpendicular to the conductor. The direction of the force is shown in Figure 1.1, and is at right angles to both the current and the magnetic field. With the magnetic field vertically downwards, and the current flowing into the paper, the force is horizontal and to the left. If either the field or the current is reversed, the force acts to the right, and if both are reversed, the force will remain to the left.

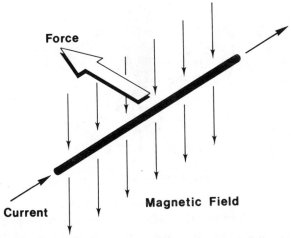

Figure 1.1   *Electromagnetic force on current-carrying conductor in a magnetic field*

These simple relationships have a pleasing feel to them: if we double either the current or the field strength, we double the force, while doubling both causes the force to increase by a factor of four. But what about quantifying the force? We need to express the force in terms of the product of the current and the magnetic field strength, and this turns out to be very straightforward when we describe the magnetic field in terms of magnetic flux density, **B**.

Magnetic flux and flux density

The familiar patterns made by iron filings in the vicinity of a bar magnet offer us clear guidance in our quest for an effective way of picturing and quantifying the magnetic field.

The tendency of the filings to form themselves into elegant curved lines, as shown in Figure 1.2, is indicative of the presence of a magnetic field; and their orientation immediately suggests the notion of a direction of the field at each point.

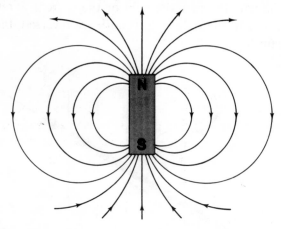

Figure 1.2   *Magnetic flux lines surrounding a bar magnet*

Experiment shows that where the field is strongest (at the ends of the magnet), the lines are close together, while in the weaker field regions (remote from the ends of the magnet), the lines are far apart. In particular, we find that if we focus on a particular pair of adjacent lines, the strength of the field halves each time the distance between the lines doubles.

If we couple these observations to the idea that between each pair of lines (and assuming a fixed distance into the paper) there is a fixed quantity of magnetic flux, we are led to the concept of magnetic flux density. When the lines are close together, the 'tube' of flux is squashed into a smaller space, giving a high flux density; whereas when the lines spread further apart the same tube of flux has more breathing space and the flux density is lower. In each case the direction of the flux density is indicated by the prevailing direction of the lines.

All that remains is to specify units for quantity of flux, and flux density. In the SI system, the unit of magnetic flux is the Weber (Wb). If one weber of flux is distributed uniformly across an area of one square metre perpendicular to the flux, the flux density is one weber per square metre, or one Tesla (T). In general if a flux (Φ) is distributed uniformly across an area A, the flux density is given by

$$B = \frac{\Phi}{A} \qquad (1.1)$$

In the motor world we are unlikely to encounter more than a few milliwebers of flux, and a small bar magnet would probably only produce a few microwebers. On the other hand, values of flux density are typically around one tesla in most motors, which is a reflection of the fact that although the quantity of flux is small, it is also spread over a small area.

Force on a conductor

We can now quantify the force, and begin to establish some feel for the magnitudes of forces we can exploit in a motor.

The force on a wire of length l, carrying a current I and exposed to a uniform magnetic flux density B is given by the simple expression

$$F = B.I.l. \qquad (1.2)$$

In equation 1.2, F is in Newtons when B is in tesla, I in amps, and l in metres. (It may come as a surprise that there are no constants of proportionality involved in equation 1.2: this is not a coincidence, but arises because the unit of current (the ampere) is actually defined in terms of force.)

Strictly, equation 1.2 only applies when the current is perpendicular to the field. If this condition is not met, the force on the conductor will be less; and in the extreme case where the current was in the same direction as the field, the force would fall to zero. However, every sensible motor designer knows that to get the best out of the magnetic field it has to be perpendicular to the conductors, and so it is safe to assume in the subsequent discussion that B and I are always perpendicular.

The reason for the very low force detected in the experiment with the bar magnet is revealed by equation 1.2. To obtain a high force, we must have a high flux density, and a lot of current. The flux density at the ends of a bar magnet is low, perhaps 0.1 tesla, so a wire carrying 1 amp will experience a force of only 0.1 N (approximately 100 gm wt) per metre. Since the flux density will be confined to perhaps 1 cm across the end face of the magnet, the total force on the wire will be only 1 gm. This would be barely detectable, and is too low to be of any use in a decent motor. So how is more force obtained?

The first step is to obtain the highest possible flux density. This is achieved by designing a good magnetic circuit, and is discussed next. Secondly, as many conductors as possible must be packed in the space where the magnetic field exists, and each conductor must carry as much current as it can without heating up to a dangerous temperature. In this way, impressive forces can be obtained from modestly sized devices, as anyone who has tried to stop an electric drill by grasping the chuck will testify.

MAGNETIC CIRCUITS

The concept of the magnetic circuit stems from the fact that magnetic flux lines always form closed contours. This is illustrated in Figure 1.3, which shows the field produced by a circular solenoid. We note that the field pattern is similar to that produced by a bar magnet, and we will see later that in many machines the source of flux can be either a coil (or winding) or a permanent magnet.

Recalling that the space between each pair of lines (and

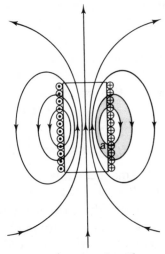

Figure 1.3 *Magnetic field of circular solenoid*

unit depth into the paper) contains a fixed quantity of flux, we can picture one tube of flux as 'starting' at, say, point a in Figure 1.3, then 'flowing' round a circuit (shown shaded) before arriving back at a.

The air surrounding the sources of the field in Figure 1.3 offers a homogeneous path for the flux, so once the tubes of flux escape from the concentrating influence of the source, they are free to spread out into the whole of the surrounding space. The flux density outside the coil is therefore low; and even inside the coil, we would find that the flux densities which we could achieve are still too low to be of use in a motor. What is needed is firstly a way of increasing the flux density, and secondly a means for preventing the flux from spreading out into the surrounding space.

Magnetomotive force (MMF)

One obvious way to increase the flux density is to increase the current in the coil, or to add more turns. We find that if we double the current, or the number of turns, we double the total flux, thereby doubling the flux density everywhere.

We quantify the ability of the coil to produce flux in terms of the Magnetomotive Force (MMF). The MMF of the coil is simply the product of the number of turns and the current, and is thus expressed in ampere-turns. A given MMF can be obtained with a large number of turns of thin wire carrying a low current, or a few turns of thick wire carrying a high current: as long as the the product NI is constant, the MMF is the same.

Electric circuit analogy and reluctance

We have seen that the magnetic flux which is set up is proportional to the MMF driving it. This points to a parallel with the electric circuit, where the current (amps) which flows is proportional to the EMF (volts) driving it.

In the electric circuit, current and EMF are related by Ohm's Law, which is

$$I = \frac{EMF}{Resistance} = \frac{V}{R}. \qquad (1.3)$$

For a given source EMF (volts), the current depends on the resistance of the circuit, so to obtain more current, the resistance of the circuit has to be reduced. We should also note that we should use low-resistance conductors to convey the current from the battery to the load, otherwise some of the source EMF will be used up in overcoming the resistance of the wires. This explains why we always use copper wires to connect source and load.

We can make use of an equivalent 'magnetic Ohm's law' by introducing the idea of Reluctance (Λ). The reluctance gives a measure of how difficult it is for the magnetic flux to complete its circuit, in the same way that resistance indicates how much opposition the current encounters in the electric circuit. The magnetic Ohm's law is then

$$Flux = \frac{MMF}{Reluctance} \quad i.e \quad \Phi = \frac{NI}{\Lambda}. \qquad (1.4)$$

We see from equation 1.4 that to increase the flux for a given MMF, we need to reduce the reluctance of the mag-

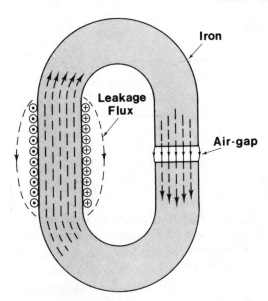

Figure 1.4 *Simple magnetic circuit*

netic circuit. In the case of the example (Figure 1.3), this means we must replace as much as possible of the air path (which is a poor magnetic material, and therefore constitutes a high reluctance) with a good magnetic material, thereby reducing the reluctance and resulting in a higher flux for a given MMF.

The material which we choose is good quality magnetic steel, which for historical reasons is usually referred to as 'iron'. This brings two desirable benefits, as shown in Figure 1.4.

Firstly, the reluctance of the iron paths is very much less than the air paths which they have replaced, so the total flux produced for a given MMF is much greater. And secondly, almost all the flux is confined within the iron, rather than spreading out into the surrounding air. We can therefore shape the iron parts of the magnetic circuit as shown in Figure 1.4 in order to guide the flux to wherever it is needed. Note also that inside the iron, the flux density remains uniform over the whole cross-section, there being so little reluctance that there is no noticeable tendency for the flux to crowd to one side or another.

The air-gap

In motors, we intend to use the high flux density to develop force on current-carrying conductors. We have now seen how to create a high flux density inside the iron parts of a magnetic circuit, but, of course, it is physically impossible to put current-carrying conductors inside the iron. We therefore arrange for an air-gap in the magnetic circuit, as shown in Figure 1.4. We will see shortly that the conductors on which the force is to be produced will be placed in this air-gap region.

If the air-gap is relatively small, as in motors, we find that the flux jumps across the air-gap as shown in Figure 1.4, without any appreciable tendency to balloon out into the surrounding air. With most of the flux lines going straight across the air-gap, the flux density in the gap region has the same high value as it does inside the iron.

In the majority of magnetic circuits consisting of iron parts and one or more air-gaps, the reluctance of the iron parts is very much less than the reluctance of the gaps. At first sight this can seem surprising, since the distance across the gap is so much less than the rest of the path through the iron. The fact that the air-gap dominates the reluctance is simply a reflection of how poor air is as a magnetic medium, compared with iron. To put the comparison in perspective, if we calculate the reluctances of two paths of equal length and cross-sectional area, one being in iron and the other in air, the reluctance of the air path will typically be 1000 times greater than the reluctance of the iron path.

Returning to the analogy with the electric circuit, the role of the iron in the magnetic circuit can be likened to that of the copper wires. Both offer little opposition to flow, and both can be shaped to guide the flow to its destination. There is one important difference, however. In the electric circuit, no current will flow until the circuit is completed, after which all the current is confined inside the wires. With an iron magnetic circuit, some flux can flow (in the surrounding air) even before the iron is installed. And although most of the flux will subsequently take the easy route through the iron, some will still leak into the air, as shown in Figure 1.4. We will not pursue leakage flux here, though it is sometimes important, as will be seen later.

Air-gap flux densities

If we neglect the reluctance of the iron parts of a magnetic circuit, it is relatively easy to estimate the flux density in the air-gap. Since the iron parts are then in effect perfect conductors of flux, none of the source MMF (NI) is used in driving the flux through the iron parts, and all of it is available to push the flux across the air-gap. The situation depicted in Figure 1.4 therefore reduces to that shown in Figure 1.5, where an MMF of NI is applied directly across an air-gap of length g.

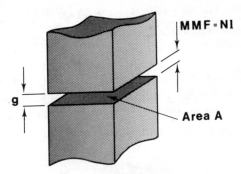

Figure 1.5 *Air-gap region, with MMF of NI applied across opposing faces*

The reluctance of any 'prism' of air, of cross-sectional area A and length g (Figure 1.5) is given by

$$\Lambda = \frac{g}{\mu_o A}. \tag{1.5}$$

where μ_o is the magnetic space constant or permeability of free space which quantifies the magnetic properties of air.

Equation 1.5 reveals the expected result that doubling the air-gap would double the reluctance (because the flux has twice as far to go), while doubling the area would halve the reluctance (because the flux has two equally appealing paths in parallel). To calculate the flux, Φ, we use the magnetic Ohm's law (equation 1.4), which gives

$$\Phi = \frac{MMF}{\Lambda} = \frac{NIA\mu_o}{g}. \tag{1.6}$$

We are usually interested in the flux density in the gap, rather than the flux, so we use equation 1.1 to yield

$$B = \frac{\Phi}{A} = \frac{\mu_o NI}{g}. \tag{1.7}$$

Equation 1.7 is very simple, and from it we can calculate the air-gap flux density once we know the MMF of the coil (NI) and the length of the gap (g). For example, suppose the magnetizing coil has 250 turns, the current is 2 A, and the gap is 1 mm. The flux density is then given by

$$B = \frac{4\pi \times 10^{-7} \times 250 \times 2}{1 \times 10^{-3}} = 0.63 \, \text{tesla}.$$

If the cross-sectional area of the iron was constant at all points, the flux density would be 0.63 T everywhere. Sometimes, however, the cross-section of the iron reduces at points away from the air-gap, as shown for example in Figure 1.6.

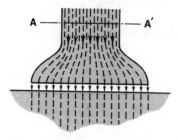

Figure 1.6   *Sketch showing how flux density varies with cross-sectional area of the magnetic circuit*

Because the flux is compressed in the narrower sections, the flux density is higher, and in Figure 1.6 if the flux density at the air-gap and in the adjacent pole-faces is once again taken to be 0.63 T, then at the section AA′ (where the area is only half that at the air-gap) the flux density will be 2 × 0.63 = 1.26 T.

Saturation

It would be reasonable to ask whether there is any limit to the flux density at which the iron can be operated. We can anticipate that there must be a limit, or else it would be possible to squash the flux into a vanishingly small cross-section, which we know from experience is not the case. In fact there is a limit, though not a very sharply defined one.

Earlier we noted that the iron has almost no reluctance, at least not in comparison with air. Unfortunately this happy state of affairs is only true as long as the flux density remains below about 1.6–1.8 T, depending on the particular steel in question. If we try to work the iron at higher flux densities, it begins to exhibit significant reluctance, and no longer behaves like an ideal conductor of flux. At these higher flux densities a significant proportion of the source MMF is used in driving the flux through the iron. This situation is obviously undesirable, since less MMF remains to drive the flux across the air-gap. So just as we would not recommend the use of high-resistance supply leads to the load in an

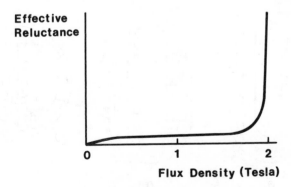

Figure 1.7 *Sketch showing how the effective reluctance of iron increases rapidly as the flux density approaches saturation*

electric circuit, we must avoid saturating the iron parts of the magnetic circuit.

The emergence of significant reluctance as the flux density is raised is illustrated qualitatively in Figure 1.7.

When the reluctance begins to be appreciable, the iron is said to be beginning to saturate. The term is apt, because if we continue increasing the MMF, or reducing the area of the iron, we will eventually reach an almost constant flux density, typically around 2 T. To avoid the undesirable effects of saturation, the size of the iron parts of the magnetic circuit are usually chosen so that the flux density does not exceed about 1.5 T. At this level of flux density, the reluctance of the iron parts will be small in comparison with the air-gap.

Magnetic circuits in motors

The reader may be wondering why so much attention has been focused on the gapped C-core magnetic circuit, when it appears to bear little resemblance to the iron parts found in motors. We shall now see that it is actually a short step from the C-core to a motor magnetic circuit, and that no fundamentally new ideas are involved.

The evolution from C-core to motor geometry is shown in Figure 1.8, which should be largely self-explanatory, and relates to the field system of a d.c. motor.

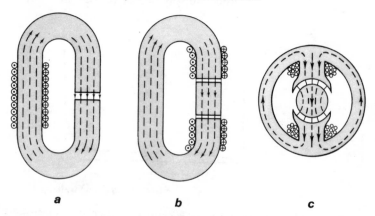

Figure 1.8 *Evolution of magnetic circuit of d.c. motor from simple C-core*

We note that the first stage of evolution (Figure 1.8(b)) results in the original single gap of length g being split into two gaps of length g/2, reflecting the requirement for the rotor to be able to turn. At the same time the single magnetizing coil is split into two to preserve symmetry. (Relocating the magnetizing coil at a different position around the magnetic circuit is of course in order, just as a battery can be placed anywhere in an electric circuit.) Finally (Figure 1.8(c)), the single magnetic path is split into two parallel paths of half the original cross-section, each of which carries half of the flux, and the pole faces are curved to match the rotor. The air-gap is still small, so the flux crosses radially to the rotor.

TORQUE PRODUCTION

Having designed the magnetic circuit to give a high flux density under the poles (Figure 1.8c), we must obtain maximum benefit from it. We therefore need to arrange a set of con-

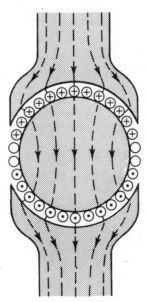

Figure 1.9   *Current-carrying conductors on rotor, positioned to maximize torque*

ductors on the rotor, and to ensure that conductors under a N-pole (at the top of Figure 1.9) carry positive current (into the paper), while those under the S-pole carry negative current. The force on all the positive conductors will be to the left, while the force on the negative ones will be to the right. A nett couple, or torque, will therefore be exerted on the rotor, which will be caused to rotate.

Magnitude of torque

The force on each conductor is given by equation 1.2, and it follows that the total tangential force F depends on the flux density produced by the field winding, the number of conductors on the rotor, the current in each, and the length of the rotor. The resultant torque (T) depends on the radius of the rotor (r), and is given by

$$T = Fr. \tag{1.8}$$

Slotting

If the conductors were mounted on the surface of the iron, as in Figure 1.9, the air-gap would have to be at least equal to the wire diameter, and the conductors would have to be secured to the rotor in order to transmit their turning force to it.

We can avoid the penalty of the large air-gap (which would result in an unwelcome high reluctance in the magnetic circuit) by placing the conductors in slots in the rotor, as shown in Figure 1.10.

With the conductors in slots, the air-gap can be made small, but, as can be seen from Figure 1.10, almost all the flux now passes down the low-reluctance path through the teeth, leaving the conductors exposed to the very low leakage flux density in the slots. It might therefore be expected that little or no force would be developed, since on the face of it the conductors are screened from the flux. Remarkably, however, what happens is that the total force remains the same as it would have been if the conductors were actually in

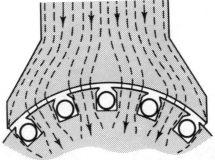

Figure 1.10   *Influence on flux paths when air-gap surface is slotted to accommodate conductors*

the flux, but almost all the force now acts on the rotor teeth, rather than on the conductors.

This is very good news indeed. By putting the conductors in slots, we simultaneously reduce the reluctance of the magnetic circuit, and transfer the force to the rotor iron, which is robust and well able to transmit the resulting torque to the shaft. There are some snags, however. To maximize the torque, we will want as much current as possible in the rotor conductors. Naturally we will work the copper at the highest practicable current density (typically between 2 and 8 A/mm^2), but we will also want to maximize the cross-sectional area of the slots to accommodate as much copper as possible. This will push us in the direction of wide slots, and hence narrow teeth. But we recall that the flux has to pass radially down the teeth, so if we make the teeth too narrow, the iron in the teeth will saturate, and lead to a poor magnetic circuit. There is also the possibility of increasing the depth of the slots, but this cannot be taken too far or the centre region of the rotor iron will become so depleted that it too will saturate.

SPECIFIC LOADINGS AND SPECIFIC OUTPUT

Specific loadings

A design compromise is inevitable in the crucial air-gap region, and designers constantly have to exercise their skills to

Plate 1.1   *Totally enclosed fan ventilated (TEFV) cage induction motor.*
This particular example is rated at 200 W at 1450 rev/min, and is at the lower
end of the power range for 3-phase versions. The case is of cast aluminium,
with cooling air provided by the external fan at the non-drive end. Note the
provision for alternative mounting (Photograph by courtesy of Brook
Crompton Parkinson Motors)

achieve the best balance between the conflicting demands on
space made by the flux (radial) and the current (axial).

As in most engineering design, guidelines emerge as to
what can be achieved in relation to particular sizes and types
of machine, and motor designers usually work in terms of
two parameters, the specific magnetic loading, and the
specific electric loading. These parameters have a direct
bearing on the output of the motor, as we will now see.

The specific magnetic loading ($\bar{B}$) is the average radial flux
density over the cylindrical surface of the rotor, while the
specific electric loading ($\bar{A}$) is the axial current per metre of
circumference on the rotor. Many factors influence the
values which can be employed in motor design, but in
essence both parameters are limited by the properties of the

materials (iron for the flux, and copper for the current), and by the cooling system employed to remove heat losses.

The specific magnetic loading does not vary greatly from one machine to another, because the saturation properties of most core steels are similar. On the other hand, quite wide variations occur in the specific electric loadings, depending on the type of cooling used.

The flow of current in the copper conductors results in heat being generated, and the current must therefore be limited to a value such that the insulation is not damaged. The more effective the cooling system, the higher the electric loading can be. For example, if the motor is totally enclosed and has no internal fan, the current density in the copper has to be much lower than in a similar motor which has a fan to provide a continuous flow of ventilating air. Similarly, windings which are fully impregnated with varnish can be worked much harder than those which are surrounded by air, because the solid body of encapsulating varnish provides a much better thermal path along which the heat can flow to the stator body. Overall size also plays a part in determining permissible electric loading, with larger motors generally having higher values than small ones.

In practice, the important point to be borne in mind is that unless an exotic cooling system is employed, most motors (induction, d.c. etc.) of a particular size have more or less the same specific loadings, regardless of type. As we shall now see, this in turn means that motors of similar size have similar torque capabilities. This fact is not widely appreciated by users, but is always worth bearing in mind.

Torque and motor volume

In the light of the earlier discussion, it follows that the tangential force per unit area of the rotor surface is equal to the product of the two specific loadings, i.e. $\bar{B}\bar{A}$. To obtain the total tangential force we must multiply by the area of the curved surface of the rotor, and to obtain the total torque we mutiply the total force by the radius.

Hence for a rotor of diameter D and length L, the total torque is given by

$$T \: \alpha \: \bar{B}\bar{A}D^2L. \tag{1.9}$$

This equation is extremely important. The term D^2L is proportional to the rotor volume, so we see that for given values of the specific magnetic and electric loadings, the torque from any motor is proportional to the rotor volume. We are at liberty to choose a long thin rotor or a short fat one, but once the volume and specific loadings are specified, we have effectively determined the torque.

It is worth stressing that we have not focused on any particular type of motor, but have approached the question of torque production from a completely general viewpoint. In essence our conclusions reflect the fact that all motors are made from iron and copper, and differ only in the way these materials are disposed. We should also acknowledge that in practice it is the overall volume of the motor which is important, rather than the volume of the rotor. But again we find that, regardless of the type of motor, there is a fairly close relationship between the overall volume and the rotor volume, for motors of similar torque. We can therefore make the bold but generally accurate statement that the overall volume of a motor is determined by the torque it has to produce. There are of course exceptions to this rule, but as a general guideline for motor selection, it is extremely useful.

Having seen that torque depends on volume, we must now turn our attention to the question of power output.

Specific output power – importance of speed

The work done by a motor delivering a torque T is equal to the torque times the angle turned through, and the power, which is the rate of working, is therefore equal to the torque times the angular speed (ω), i.e.

$$P = T\omega. \tag{1.10}$$

We can now express the power output in terms of the rotor dimensions and the specific loadings, using equation 1.9 which yields

$$P \ \alpha \ \bar{B}\bar{A}D^2L\omega. \tag{1.11}$$

Equations 1.10 and 1.11 emphasize the importance of speed in determining power output. For a given power, we can choose between a large low-speed motor or a small high-speed one. The latter choice is preferred for most applications, even if some form of speed reduction (using belts or gears, for example) is needed. Familiar examples include portable electric tools, where rotor speeds of 12,000 rev/min or more allow powers of hundreds of watts to be obtained, and electric traction, where the motor speed is considerably higher than the wheel speed. In these examples, volume and weight are at a premium, and a direct drive would be out of the question.

The significance of speed is underlined when we rearrange equation 1.11 to obtain an expression for the specific power output (power per unit rotor volume), Q, given by

$$Q \ \alpha \ \bar{B}\bar{A}\omega. \tag{1.12}$$

To obtain the highest possible specific output for given values of the specific magnetic and electric loadings, we must clearly operate the motor at the highest practicable speed.

MOTIONAL EMF

We have already seen that force (and hence torque) is produced on current-carrying conductors exposed to a magnetic field. The force is given by equation 1.2, which shows that as long as the flux density and current remain constant, the force will be constant. In particular we see that the force does not depend on whether the conductor is stationary or moving. On the other hand relative movement is an essential requirement in the production of mechanical output power (as distinct from torque), and we have seen that output power is given by the equation $P = T\omega$. We will now see that the presence of relative motion between the conductors and the field always brings 'motional e.m.f.' into play; and we shall see that this motional e.m.f. is an essential feature of the energy conversion process.

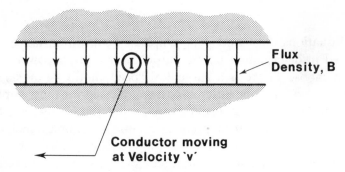

Figure 1.11   *Elementary 'motor' showing conductor carrying current I, and moving with velocity V perpendicular to magnetic field*

Power relationships – stationary conditions

We look first at the electrical input power, considering the linear set-up depicted in Figure 1.11, and beginning with the conductor stationary (i.e. v = 0).

For the purpose of this discussion we can suppose that the magnetic field (B) is provided by a permanent magnet. Once the field has been established (when the magnet was first magnetized and placed in position), no further energy will be needed to sustain the field. It should therefore be obvious that when we obtain mechanical output from this primitive motor, none of the energy involved comes from the magnet.

The only electrical input power required when the conductor is held stationary is that needed to drive the current through the conductor. If the resistance of the conductor is R, and the current through it is I, the voltage required will be given by $V_1 = IR$, and the electrical input power will be $V_1 I$ or $I^2 R$. Force ($= BII$) is produced on the conductor, but no work is done because the conductor is being held stationary, and hence there is no mechanical power output. All the electrical input power will therefore appear as heat inside the conductor. In this condition the power balance can be expressed by the equation

Electrical Input Power $(V_1 I) =$ (1.13)
            Heat Lost in Conductor $(I^2 R)$.

Power relationships – conductor moving at velocity v

Now let us imagine the situation where the conductor is allowed to move at a constant velocity (v) in the direction of the force, and that the current in the conductor is maintained at the value (I) which it had when it was stationary. We need not worry at this stage about the practicalities of being able to supply current to a moving conductor; the important point is that the situation is typical of what happens in a real machine.

Mechanical work is now being done against the opposing force of the load, and the mechanical output power is equal to the rate of work, i.e. the force (BIl) times the velocity (v). The power lost as heat in the conductor is the same as it was when stationary, since it has the same resistance, and the same current. The electrical input power must continue to supply this heat loss, but in addition it must now furnish the mechanical output power. The power balance equation now becomes

Electrical Input Power = Heat Loss +
$$\text{Mechanical Output Power}$$

or

$$\text{Electrical Input Power} = I^2R + BIlv. \qquad (1.14)$$

Clearly, we would expect the electrical input power to be greater than it was when the conductor was stationary, and we can see from equation 1.14 that it must increase directly with the speed. The question which then springs to mind is how does the source supplying current to the conductor know that it has to supply more power in proportion to the speed of the conductor? There must be some electrical effect which accompanies the motion, and causes a change in the way the conductor appears to the supply, otherwise the situation on the electrical side would be unchanged.

The key to unlock the question lies in our assumption that the current is constant all the time. Electrical input power is equal to current times voltage, so since the current is assumed to have been kept constant it follows that the voltage of the source must increase from the value it had when the conductor was stationary, by an amount sufficient to furnish the extra mechanical power. If we denote the required source voltage when the conductor is moving by V_2, the power balance equation can be written

$$V_2 I = I^2 R + BIlv. \qquad (1.15)$$

Combining equations 1.13 and 1.15 we obtain

$$(V_2 - V_1)I = BlvI \qquad (1.16a)$$

and thus

$$V_2 - V_1 = E = Blv. \qquad (1.16b)$$

The first of these equations shows the expected result that the increase in source power is equal to the mechanical output power, while the second quantifies the extra voltage to be provided by the source to keep the current constant when the conductor is moving. This increase in source voltage is a reflection of the fact that when a conductor moves through a magnetic field, an e.m.f (E) is induced in it.

We see from equation 1.16b that the e.m.f. is directly proportional to the flux density, to the velocity of the conductor relative to the flux, and to the length of the conductor. The source voltage has to overcome this additional voltage in order to keep the same current flowing: if the source voltage was not increased, the current would fall as soon as the conductor began to move because of the opposing effect of the induced e.m.f.

We have deduced that there must be an e.m.f. caused by the motion, and have derived an expression for it by using the principle of the conservation of energy, but the result we have obtained, i.e.

$$E = BLv \qquad (1.17)$$

is often introduced as the 'flux-cutting' form of Faraday's law, which states that when a conductor moves through a magnetic field an e.m.f. given by equation 1.17 is induced in it. Because motion is an essential part of this mechanism, the e.m.f. induced is referred to as 'motional e.m.f.'. The 'flux-cutting' terminology arises from attributing the origin of the e.m.f. to the cutting or slicing of the lines of flux by the passage of the conductor. This is a useful mental picture, though it must not be pushed too far: the flux lines are after all merely inventions which we find helpful in coming to grips with magnetic matters.

Before turning to the equivalent circuit two general points are worth noting. Firstly, whenever energy is being converted from electrical to mechanical form, as here, the induced e.m.f. always acts in opposition to the applied (source) voltage. This is reflected in the use of the term 'back e.m.f.' to describe motional e.m.f. in motors. Secondly, although we have discussed a particular situation in which the conductor carries current, it is certainly not necessary for any current to be flowing in order to produce an e.m.f. All that is needed is relative motion between the conductor and the magnetic field.

Equivalent circuit

We can represent the electrical relationships in the moving conductor experiment in an equivalent circuit as shown in Figure 1.12.

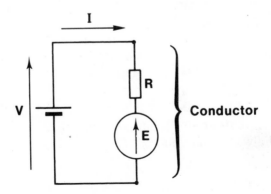

Figure 1.12   *Equivalent circuit for the elementary motor shown in Figure 1.11*

The motional e.m.f. in Figure 1.12 is shown as opposing the applied voltage, which applies in the 'motoring' condition we have been discussing. Applying Kirchoff's law we obtain the voltage equation as

$$V = E + IR \quad \text{or} \quad I = \frac{V - E}{R}. \qquad (1.18)$$

Multiplying equation 1.18 by the current gives the power equation as

Electrical Input Power (VI) =
Mechanical Output Power (EI) + Copper Loss (I^2R).   (1.19)

It is worth seeing what can be learned from these equations because, as noted earlier, this simple set-up encapsulates all the essential features of real motors. Lessons which emerge at this stage will be invaluable later, when we look at the way actual motors behave.

If the e.m.f. E is less than the applied voltage V, the current will be positive, and electrical power will flow from the source, resulting in motoring action. On the other hand if E is larger than V, the current will flow back to the source, and the conductor will be acting as a generator. This inherent ability to switch from motoring to generating without any interference by the user is an extremely desirable property of electromagnetic energy converters. Our primitive set-up is simply a machine, which is equally at home acting as motor or generator.

A further important point to note is that the mechanical power (the first term on the right hand side of equation 1.19) is simply the motional e.m.f. multiplied by the current. This result is again universally applicable, and easily remembered. We may sometimes have to be a bit careful if the e.m.f. and the current are not simple d.c. quantities, but the basic idea will always hold good.

Motoring condition

Motoring implies that the conductor is moving in the same direction as the electromagnetic force (BIl), and at a speed such that the back e.m.f. (BLv) is less than the applied voltage V. In the discussion so far, we have assumed that the applied voltage is adjusted so that the current is kept constant. This was a helpful approach to take in order to derive the steady-state power relationships, but is seldom typical of normal operation. We therefore turn to how the moving conductor will behave under conditions where the applied voltage V is constant, since this corresponds more closely with operation of a real motor.

Behaviour without load

If we begin with the conductor stationary when the voltage V is first applied, the current will immediately rise to a value of V/R, since there is no motional e.m.f. and the only thing which limits the current is the resistance. (Strictly we should allow for the effect of inductance in delaying the rise of current, but we choose to ignore it here in the interests of simplicity.) The current will be large, and a high force will therefore be developed. If there are no other forces acting, the conductor will accelerate in the direction of the force. As it picks up speed, the motional e.m.f. will grow in proportion to the speed. Since the motional e.m.f. opposes the applied voltage, the current will fall (Equation 1.18), so the force and hence the acceleration will reduce. The speed will thus rise exponentially, until it reaches an equilibrium condition where the back e.m.f. is equal to the applied voltage, and the current has fallen to zero, as shown in Figure 1.13. The conductor will then continue to travel at a constant speed, with no nett force acting on it. No mechanical power is being produced in the steady state, since we have assumed that there is no opposing force on the conductor. This situation corresponds to the so-called no-load condition in a motor, the only difference being that a motor will have some friction, which has been ignored here in order to simplify the discussion.

An elegant self-regulating mechanism is evidently at work here. When the conductor is stationary, it has a high force

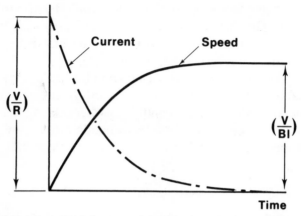

Figure 1.13   *Dynamic behaviour of the elementary motor with no mechanical load*

acting on it, but this force tapers-off as the speed rises to its target value, which corresponds to the back e.m.f. being equal to the applied voltage. Looking back at the expression for motional e.m.f. (equation 1.17), we can obtain an expression for the no-load speed (v_0) by equating the applied voltage and the back e.m.f., which gives

$$v_o = \frac{E}{Bl} = \frac{V}{Bl}. \tag{1.20}$$

Equation 1.20 shows that the steady-state no-load speed is directly proportional to the applied voltage, which indicates that speed control can be achieved by means of the applied voltage.

Rather more surprisingly, however, the speed is seen to be inversely proportional to the magnetic flux density, which means that the weaker the field, the higher the steady-state speed. This result can cause raised eyebrows, and with good reason. Surely, it is argued, since the force is produced by the action of the field, the conductor will not go as fast if the field is weaker. This view is wrong, but understandable. The flaw in the argument is to equate force with speed. When the voltage is first applied, the force on the conductor certainly will be less if the field is weaker, and the initial acceleration

will be lower. But in both cases the acceleration will continue until the current has fallen to zero, and this will only
happen when the induced e.m.f. has risen to equal the
applied voltage. With a weaker field, the speed needed to
generate this e.m.f. will be higher than with a strong field:
there is 'less flux', so what there is has to be cut at a higher
speed to generate a given e.m.f. The matter is summarized in
Figure 1.14, which shows how the speed will rise for a given
applied voltage, for 'full' and 'half' fields respectively.

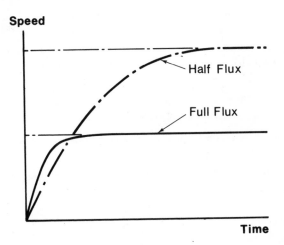

Figure 1.14   *Effect of flux on the acceleration and steady running speed of
the elementary motor*

Behaviour when loaded

Suppose we now apply a steady force (the load) opposing the
motion of the conductor. The conductor will begin to decelerate, but as soon as the speed falls, the back e.m.f. will
fall to a value less than V, and current will begin to flow.
The more the speed drops, the bigger the current, and hence
the more the electromagnetic force developed by the conductor. When the force developed by the conductor equals
the force we have applied, the deceleration will cease, and a
new equilibrium condition will be reached. The speed will
be lower than at no-load, and the conductor will now be

producing continuous mechanical output power, i.e. it will be acting as a motor.

Since the electromagnetic force on the conductor is directly proportional to the current, it follows that the steady-state current is directly proportional to the load which is applied. We note from equation 1.18 that the current depends directly on the difference between V and E, and inversely on the resistance. Hence for a given resistance, the larger the steady-state current, the smaller E has to be for a given applied voltage V. Since E is proportional to speed, we see that the drop in speed for a given current (i.e. a given load) is directly proportional to the resistance. Hence the lower the resistance, the more the conductor is able to hold its no-load speed in the face of applied load. This is illustrated in Figure 1.15, from which we see that a high resistance leads to a large drop in speed for a given applied load, while a low resistance leads to much better speed holding.

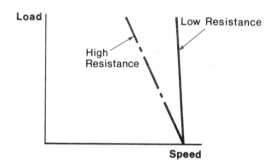

Figure 1.15   *Influence of resistance on the drop in speed with load*

We complete our exploration of the performance when loaded by asking how the flux density influences behaviour. Recalling that the electromagnetic force is proportional to the flux density as well as the current, we can deduce that to develop a given force, the current required will be higher with a weak flux than with a strong one. Hence in view of the fact that there will always be an upper limit to the current which the conductor can safely carry, the maximum force which can be developed will vary in direct proportion to the flux density, so a weak flux will lead to a low maximum force and vice-versa. In addition, to achieve any given force, the drop in speed will be disproportionately high when we go to a lower flux density. We can see this by imagining that we want a particular force, and considering how we achieve it firstly with full flux, and secondly with half flux.

With full flux, there will be a certain drop in speed which causes the motional e.m.f. to fall enough to admit the required current. With half the flux, twice as much current will be needed to develop the same force. Hence the motional e.m.f. must fall by twice as much as it did with full flux. However, since the flux density is now only half, the drop in speed will have to be four times as great as it was previously. The half-flux 'motor' therefore has a much more droopy load charactersitic than the full-flux one. This is shown in Figure 1.16, the applied voltages having been adjusted so that in both cases the no-load speed is the same.

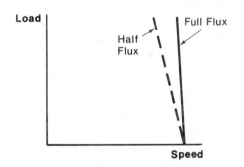

Figure 1.16   *Influence of flux on the drop in speed with load*

We may be tempted to think that the higher speed which we can obtain by reducing the flux somehow makes for better performance, but we can now see that this is not so. By halving the flux, for example, the no load speed for a given voltage is doubled, but when the load is raised until rated current is flowing in the conductor, the force developed is only half, so the mechanical power is the same. We are in

effect trading speed against force, and there is no suggestion
of getting something for nothing.

Relative magnitudes of V, E and efficiency

Invariably we want machines which have high efficiency.
From equation 1.19, we see that to achieve high efficiency, the
copper loss (I^2R) must be small compared with the mechanical power (EI), which means that the resistive volt-drop in
the conductor (IR) must be small compared with either the
induced e.m.f. (E) or the applied voltage (V). In other words
we want most of the applied voltage to be accounted for by
the useful motional e.m.f., rather than the wasteful volt
drop in the wire. Since the motional e.m.f. is proportional to
speed, and the resistive volt drop depends on the conductor
resistance, we see that a good energy converter requires the
conductor resistance to be as low as possible, and the speed
to be as high as possible.

To provide a feel for the sorts of numbers likely to be
encountered, we can consider a conductor with resistance of
$0.5\,\Omega$, carrying a current of 4 A, and moving at a speed such
that the motional e.m.f. is 8 V. From equation 1.18, the
supply voltage is given by

$$V = E + IR = 8 + 4 \times 0.5 = 10\,\text{volts.} \qquad (1.21)$$

Hence the electrical input power (VI) is 40 watts, the mechanical output power (EI) is 32 watts, and the copper loss
(I^2R) is 8 watts, giving an efficiency of 80 per cent.

If the supply voltage was doubled, however, and the
resisting force is assumed to remain the same (so that the
steady-state current is still 4 A), the motional e.m.f. is given
by equation 1.18 as

$$E = 20 - 4 \times 0.5 = 18\,\text{volts} \qquad (1.22)$$

which shows that the speed will have rather more than
doubled, as expected. The electrical input power is now 80
watts, the mechanical output power is 72 watts, and the
copper loss is still 8 watts. The efficiency has now risen to 90
per cent, underlining the fact that the energy conversion process gets better at higher speeds.

The ideal situation is clearly one where the term IR in equation 1.18 is negligible, so that the back e.m.f. is equal to the applied voltage. We would then have an ideal machine with an efficiency of 100 per cent, in which the steady-state speed would be directly proportional to the applied voltage and independent of the load.

In practice the extent to which we can approach the ideal situation discussed above depends on the size of the machine. Tiny motors, such as those used in wrist watches, are awful, in that most of the applied voltage is used up in overcoming the resistance of the conductors, and the motional e.m.f. is very small: these motors are much better at producing heat than they are at producing mechanical output power! Small machines, such as those used in hand tools, are a good deal better with the motional e.m.f. accounting for perhaps 70–80% of the applied voltage. Industrial machines are very much better: the largest ones (of many hundreds of kW) use only one or two per cent of the applied voltage in over-coming resistance, and therefore have very high efficiencies.

GENERAL PROPERTIES OF ELECTRIC MOTORS

All electric motors are governed by the laws of electromagnetism, and are subject to essentially the same constraints imposed by the materials (copper and iron) from which they are made. We should therefore not be surprised to find that at the fundamental level all motors – regardless of type – have a great deal in common.

These common properties, most of which have been touched on in this chapter, are not usually given prominence. Books tend to concentrate on the differences between types of motor, and manufacturers are usually interested in promoting the virtues of their particular motor at the expense of the competition. This divisive emphasis causes the underlying unity to be obscured, leaving users with little opportunity to absorb the sort of knowledge which will equip them to make informed judgements.

The most useful ideas worth bearing in mind are therefore

Plate 1.2   *Steel frame cage induction motor, 150 kW, 1485 rev/min. The active parts are totally enclosed, and cooling is provided by means of an internal fan which circulates cooling air round the interior of the motor through the hollow ribs, and an external fan which blows air over the case (Photograph by courtesy of Brook Crompton Parkinson Motors)*

given below, with brief notes accompanying each. Experience indicates that users who have these basic ideas firmly in mind will find themselves able to understand why one motor seems better than another, and will feel much more confident when faced with the difficult task of weighing the pros and cons of competing types.

Operating temperature and cooling

The cooling arrangement is the single most important factor in determining the output from any given motor.

Any motor will give out more power if its electric circuit is worked harder (i.e. if the current is allowed to increase). The limiting factor is normally the allowable temperature rise of the windings, which depends on the class of insulation.

For class F insulation (the most widely used) the permissible temperature rise is 100 K, whereas for class H it is 125 K. Thus if the cooling remains the same, more output can be obtained simply by using the higher-grade insulation. Alternatively, with a given insulation the ouput can be increased if the cooling system is improved. A through-ventilated motor, for example, might give perhaps twice the output power of an otherwise identical but totally enclosed machine.

Torque per unit volume

For motors with similar cooling systems, the torque per unit volume is approximately proportional to the rotor volume, which in turn is roughly proportional to the overall motor volume.

This stems from the fact that for a given cooling arrangement, the specific and magnetic loadings of machines of different types will be more or less the same. The torque per unit length therefore depends first and foremost on the square of the diameter, so motors of roughly the same diameter and length can be expected to produce roughly the same torque.

Power per unit volume – importance of speed

Output power per unit volume is directly proportional to speed.

Low-speed motors are unattractive because they are large. It is usually much better to use a high-speed motor with a mechanical speed reduction. For example, a direct drive motor for a portable electric screwdriver would be an absurd proposition.

Size effects – specific torque and efficiency

Large motors have a higher specific torque (torque per unit volume) and are more efficient than small ones.

In large motors the specific electric loading is normally

much higher than in small ones, and the specific magnetic loading is somewhat higher. These two factors combine to give the higher specific torque.

Very small motors are inherently very inefficient (e.g. 1 per cent in a wrist-watch), whereas motors of over say 100 kW have efficiencies above 95 per cent. The reasons for this scale effect are complex, but stem from the fact that the resistance volt-drop term can be made relatively small in large electromagnetic devices, whereas in small ones the resistance becomes the dominant term.

Efficiency and speed

The efficiency of a motor improves with speed.

For a given torque, power output rises in proportion to speed, while electrical losses are – broadly speaking – constant. Efficiency therefore rises with speed.

Rated voltage

A motor can be provided to suit any voltage.

Within limits it is always possible to rewind a motor for a different voltage without affecting its performance. A 200 V, 10 A motor could be rewound for 100 V, 20 A simply by using half as many turns per coil of wire having twice the cross-sectional area. The total amounts of active material, and hence the performance, would be the same.

Short-term overload

Most motors can be overloaded for short periods without damage.

The continuous electric loading (i.e. the current) cannot be exceeded without damaging the insulation, but if the motor has been running with reduced current for some time, it is permissible for the current (and hence the torque) to be much greater than normal for a short period of time. The principal factors which influence the magnitude and duration of the permissible overload are the thermal time-constant (which governs the rate of rise of temperature) and the previous pattern of operation. Thermal time constants range from a few seconds for small motors to many minutes or even hours for large ones. Operating patterns are obviously very variable, so rather than rely on a particular pattern being followed, it is usual for motors to be provided with over-temperature protective devices (e.g. thermistors) which trigger an alarm and/or trip the supply if the safe temperature is exceeded.

2

POWER ELECTRONIC CONVERTERS
FOR MOTOR DRIVES

INTRODUCTION

In this chapter we will look at examples of the power converter circuits which are used with motor drives, providing either d.c. or a.c. outputs, and working from either a d.c. (battery) supply, or from the conventional a.c. mains. The treatment is not intended to be exhaustive, but should serve to highlight the most important aspects which are common to all types of drive converter.

Although there are many different types of converter, all except very low power ones are based on some form of electronic switching. The need to adopt a switching strategy is emphasized in the first example, where the consequences are explored in some depth. We shall see that switching is essential in order to achieve high-efficiency power conversion, but that the resulting waveforms are inevitably less than ideal from the point of view of the motor.

The examples have been chosen to illustrate current practice, so for each the most commonly used switching devices are shown. More than one switching device may be suitable, however, so we should not identify a particular circuit as being the exclusive province of a particular device.

Before discussing particular circuits it will be useful to take an overall look at a typical drive system, so that the role of the converter can be seen in its proper context.

General arrangement of drive

A complete drive system is shown in block diagram form in
Figure 2.1

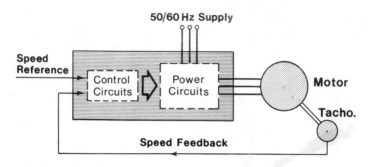

Figure 2.1 *General arrangement of speed-controlled drive*

The job of the converter is to draw electrical energy from the
mains (at constant voltage and frequency) and supply electrical energy to the motor at whatever voltage and frequency
is necessary to achieve the desired mechanical output.

Except in the very simplest converter (such as a simple
diode rectifier), there are usually two distinct parts to the
converter. The first is the power stage, through which the
energy flows to the motor, and the second is the control
section, which regulates the power flow. Control signals, in
the form of low-power analogue or digital voltages, tell the
converter what it is supposed to be doing, while other low-
power feedback signals are used to measure what is actually
happening. By comparing the demand and feedback signals,
and adjusting the output accordingly, the target output is
maintained. The simple arrangement shown in Figure 2.1 has
only one input representing the desired speed, and one feedback signal indicating actual speed, but most drives will
have extra feedback signals as we will see later.

A characteristic of power electronic converters which is
shared with most electrical systems is that they have very
little capacity for storing energy. This means that any sudden
change in the power supplied by the converter to the motor

must be reflected in a sudden increase in the power drawn from the supply. In most cases this is not a serious problem, but it does have two drawbacks. Firstly, sudden changes in the current drawn from the supply will cause spikes in the supply voltage because of the effect of the supply impedance. These spikes will appear as unwelcome distortion to other users on the same supply. And secondly, there may be an enforced delay before the supply can react. With a single-phase mains supply, for example, there can be no sudden increase in the power supply from the mains at the instant where the mains voltage is passing through zero.

It would be better if a significant amount of energy could be stored within the converter itself: short-term energy demands could then be met instantly, thereby reducing rapid fluctuations in the power drawn from the mains. But unfortunately this is just not economic: most converters do have a small store of energy in their smoothing inductors and capacitors, but the amount is not sufficient to buffer the supply sufficiently to shield it from anything more than very short-term fluctuations.

VOLTAGE CONTROL – D.C. OUTPUT FROM D.C. SUPPLY

For the sake of simplicity we will begin with the problem of controlling the voltage across a resistive load, fed from a battery. Three different methods are shown in Figure 2.2. The battery voltage is assumed to be constant at 12 V, and we seek to vary the load voltage from 0 to 12 V. Although this is not quite the same as if the load was a d.c. motor the conclusions which we draw are effectively the same.

Method (a) uses a variable resistor (R) to absorb whatever fraction of the battery voltage is not required at the load. It provides smooth control, but the snag is that power is wasted in the control resistor. For example, if the load voltage is to be reduced to 6 V, the resistor R must be set to $2\,\Omega$, so that half of the battery voltage is dropped across R. The current

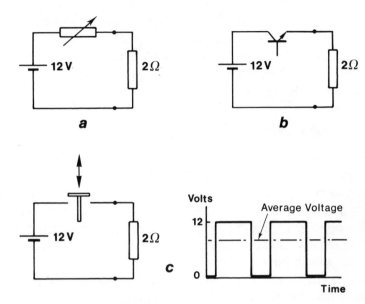

Figure 2.2 *Methods of obtaining variable-voltage d.c. output*

will be 3 A, the load power will be 18 W, and the power dissipated in R will also be 18 W, giving an overall efficiency of only 50 per cent. If R is increased further, the efficiency falls still lower, approaching zero as the load voltage tends to zero. This method of control is therefore unacceptable for motor control, except perhaps in low power applications such as car wiper and heater motors.

Method (b) is much the same as (a) except that a transistor is used instead of a manually-operated variable resistor. A transistor is a variable resistor, of course; but one in which the collector-emitter resistance can be controlled over a wide range by means of the base-emitter current. (The base-emitter current is usually very small, so it can be varied by means of a low-power electronic circuit whose losses are negligible in comparison with the power in the main (collector-emitter) circuit.)

The drawback of method (b) is the same as in (a) above, i.e. the efficiency is very low. But here the wasted power (up to a maximum of 18 W in the example) is burned-off inside the transistor which therefore has to be large, well-cooled, and hence expensive. Transistors are hardly ever operated in this linear way when used in power electronics, but are widely used as switches, as discussed below.

Switching control

The basic ideas underlying a switching power regulator are shown by the arrangement in Figure 2.2(c), which uses a mechanical switch. By operating the switch repetitively and varying the ratio of 'on to off' time, the average load voltage can be varied smoothly between 0 V (switch off all the time) through 6 V (switch 'on and off' for half of each cycle) to 12 V (switch on all the time).

The circuit shown in Figure 2.2(c) is often referred to as a 'chopper', because the battery supply is chopped 'on and off'. When a constant repetition frequency is used, and the width of the on pulse is varied to control the mean output voltage, the arrangement is known as 'pulse width modulation' (PWM). An alternative approach is to keep the width of the on pulses constant, but vary their repetition rate, and this is known as pulse frequency modulation.

The main advantage of the chopper circuit is that no power is wasted, and the efficiency is thus 100 per cent. When the switch is on, current flows through it, but the voltage across it is zero because its resistance is negligible. The power dissipated in the switch is therefore zero. Likewise, when 'off', the current is zero, so although the voltage across the switch is 12 V, the power dissipated in it is again zero.

The disadvantage is that the load voltage waveform is no longer steady: it consists of a mean 'd.c.' level, with a superimposed 'a.c.' component. Bearing in mind that we really want the load to be a d.c. motor, rather than a resistor, we are bound to ask whether the pulsating voltage will be acceptable. The answer is yes, provided the frequency is high enough. If we pulse the switch at too low a rate, we will find that the speed of the motor will fluctuate in sympathy. But as the pulse rate is increased (keeping the same mark-space ratio), the speed ripple will reduce until it is imperceptible.

Obviously a mechanical switch would be inconvenient, and could not be expected to last long when pulsed at high frequency. So a pulsed transistor switch is used instead.

Transistor chopper

As noted earlier, a transistor is effectively a controllable resistor, i.e. the resistance between collector and emitter depends on the current in the base-emitter junction. In order to mimic the operation of a mechanical switch, the transistor would have to be able to provide infinite resistance (corresponding to an open switch) or zero resistance (corresponding to a closed switch). Neither of these ideal states can be reached with a real transistor, but both can be closely approximated.

The transistor will be 'off' when the base-emitter current is zero. Viewed from the main (collector-emitter) circuit, its resistance will be very high, as shown by the region Oa in Figure 2.3.

Under this 'cut-off' condition, only a tiny current will flow

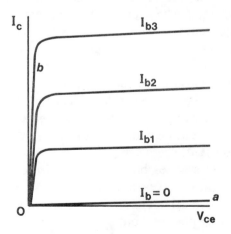

Figure 2.3   *Transistor characteristics showing high resistance (cut-off) region Oa, and low resistance (saturation) region Ob*

from collector to emitter, and the power dissipated will be negligible, giving an excellent approximation to an open switch.

To turn the transistor fully 'on', a base-emitter current must be provided. The base current required will depend on the prospective collector-emitter current, i.e. the current in the load. The aim is to keep the transistor 'saturated' so that it has a very low resistance, corresponding to the region Ob in Figure 2.3. Typically the base current will need to be perhaps 5 per cent of the collector current. By keeping the 'on' transistor in the saturation region its effective resistance will remain low for quite high collector currents. In the example, with the full load current of 6 A flowing, the collector-emitter voltage might be say 0.33 V, giving an on-state dissipation of 2 W in the transistor when the load power is 72 W. This is not as good as a mechanical switch, but is acceptable.

Just as we have to select mechanical switches with regard to their duty, we must be careful to use the right transistor for the job in hand. In particular, we need to ensure that when the transistor is 'on', we don't exceed the safe current, or else the active region of the device will be destroyed by overheating. And we must make sure that the transistor is able to withstand whatever voltage appears across the collector-emitter junction when it is in the 'off' condition. If the safe voltage is exceeded, the transistor will break down, and be permanently 'on'.

A suitable heatsink will be a necessity. We have already seen that some heat is generated when the transistor is on, and at low switching rates this is the main source of unwanted heat. But at high switching rates, 'switching loss' can be very important.

Switching loss arises because the transition from 'on' to 'off' or vice-versa takes a finite time. We will of course arrange the base-drive circuitry so that the switching takes place as fast as possible, but in practice it will seldom take less than a few microseconds. During the switch-on period, for example, the current will be building up, while the collector-emitter voltage will be falling towards zero. The peak power reached can therefore be large, before falling to the relatively low on-state value. Of course the total energy released as heat each time the device switches is modest because the whole process happens so quickly. Hence if the swiching rate is low (say once every second) the switching power loss will be insignificant in comparison with the on-state power. At high switching rates, (say 100 kHz) however, when the time taken to complete the switching becomes comparable with the on time, the switching power loss can easily become dominant. In practice, converters used in drives rarely employ switching rates much above 20 kHz, in order to minimize switching losses.

Chopper with inductive load – overvoltage protection

So far we have looked at chopper control of a resistive load, but in a drives context the load will usually mean the winding of a machine, which will invariably be inductive.

Chopper control of inductive loads is much the same as for resistive loads, but we have to be careful to prevent the appearance of dangerously high voltages each time the inductive load is switched 'off'. The root of the problem lies with the energy stored in the associated magnetic field. When an inductance L carries a current I, the energy stored in the magnetic field (W) is given by

$$W = \frac{1}{2}LI^2. \tag{2.1}$$

If the inductor is supplied via a mechanical switch, and we try to open the switch with the intention of reducing the current to zero instantaneously, we are in effect seeking to destroy the stored energy. This is not possible, and what happens is that the energy is dissipated in the form of a spark across the contacts of the switch. This sparking will be familiar to anyone who has pulled off the low-voltage lead to the ignition coil in a car: serious sparking will also be observed across the contact-breaker points themselves whenever the spark suppression capacitor is faulty.

The appearance of a spark indicates that there is a very high voltage which is sufficient to break down the surrounding air. We can anticipate this by remembering that the voltage and current in an inductance are related by the equation

$$V = L\frac{dI}{dt}. \qquad (2.2)$$

The voltage is proportional to the rate of change of current, so when we open the switch in order to force the current to zero quickly, a very large voltage is created in the inductance. This voltage appears across the switch, and if sufficient to break down the air, the current can continue to flow in the form of an arc.

Sparking across a mechanical switch is unlikely to cause immediate destruction, but when a transistor is used sudden death is certain unless steps are taken to tame the stored energy. The usual remedy lies in the use of a 'freewheel diode', as shown in Figure 2.4.

A diode is a one-way valve as far as current is concerned: it offers very little resistance to current flowing from anode to cathode, but blocks current flow from cathode to anode. Hence in the circuit of Figure 2.4(a), when the transistor is on, current (I) flows through the load, but not through the diode, which is said to be reverse-biased (i.e. the applied voltage is trying – unsuccessfully – to push current down through the diode).

When the transistor is turned off, the current through it and the battery drops very quickly to zero. But the stored energy in the inductance means that its current cannot suddenly disappear. So since there is no longer a path through the transistor, the current diverts into the only other route available, and flows upwards through the low-resistance path offered by the diode, as shown in Figure 2.4(b).

Obviously the current no longer has a battery to drive it,

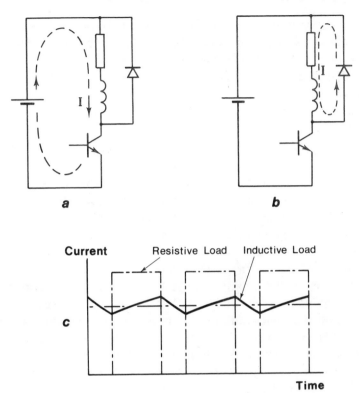

Figure 2.4   *Operation of chopper-type regulator*

so it cannot continue to flow indefinitely. In fact it will continue to freewheel only until the energy originally stored in the inductance is dissipated as heat, mainly in the load resistance but also in the diode's own (low) resistance. The current waveform during chopping will then be as shown in Figure 2.4(c). Note that the current rises and falls exponentially, with a time-constant of L/R, and that the presence of inductance causes the current to be much smoother than with a purely resistive load.

Finally, we need to check the effect of the freewheel diode on the voltage across the transistor. When a diode conducts, the forward-bias volt-drop across it is small – typically 0.7 volts. Hence while the current is freewheeling, the voltage at the collector of the transistor is only 0.7 volts above the battery voltage. This 'clamping' action therefore limits the voltage across the transistor to a safe value, and allows inductive loads to be switched without damage to the switching element.

Features of power electronic converters

We can draw some important conclusions which are valid for all power electronic converters from this simple example. Firstly, efficient control of voltage (and hence power) is only feasible if a switching strategy is adopted. The load is alternately connected and disconnected from the supply by means of an electronic switch, and any average value up to the supply voltage can be obtained by varying the mark/space ratio. Secondly, the output voltage is not smooth d.c., but contains unwanted a.c. components which, though undesirable, are tolerable in motor drives. And finally, the load current waveform will be smoother than the voltage waveform if – as is the case with motor windings – the load is inductive.

D.C. FROM A.C. – CONTROLLED RECTIFICATION

The vast majority of drives draw their power from constant voltage 50 Hz or 60 Hz mains, and in nearly all mains converters the first stage consists of a rectifier which converts the a.c. to a crude form of d.c. Where a constant-voltage d.c. output is required, a simple (uncontrolled) diode rectifier is sufficient. But where the mean d.c. voltage has to be controllable (as in a d.c. motor drive), a controlled rectifier is used.

Many different converter configurations based on combinations of diodes and thyristors are possible, but we shall focus on 'fully-controlled' converters in which all the rectifying devices are thyristors. These are especially important in the context of modern motor drives.

From the user's viewpoint, interest centres on the following questions:

- How is the output voltage controlled?
- What does the converter output voltage look like? Will there be any problems if the voltage is not pure d.c?
- How does the range of the output voltage relate to the a.c. mains voltage?

We can answer these questions without going too thoroughly into the detailed workings of the converter. This is just as well, because understanding all the ins and outs of converter operation is beyond our scope. On the other hand it is well worth trying to understand the essence of the controlled rectification process, because it assists in understanding the limitations which the converter puts on drive performance (see Chapter 4). Before tackling the questions posed above, however, it is obviously necessary to introduce the thyristor.

The thyristor and controlled rectification

The thyristor is an electronic switch, with two main terminals (anode and cathode) and a 'switch-on' terminal (gate), as shown in Figure 2.5. Like a diode, current can only flow in the forward direction, from anode to cathode. But unlike a diode, which will conduct in the forward direction as soon as forward voltage is applied, the thyristor will continue to block forward current until a small current pulse is injected into the gate-cathode circuit, to turn it on or 'fire' it. After the gate pulse is applied, the main anode-cathode current builds up rapidly, and as soon as it reaches the 'latching' level, the gate pulse can be removed and the device will remain on.

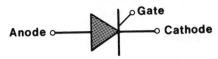

Figure 2.5   *Circuit diagram of thyristor showing anode, cathode and gate terminals*

Once established, the anode-cathode current cannot be interrupted by any gate signal. The non-conducting state can only be restored after the anode-cathode current has reduced to zero, and has remained at zero for the turn-off time (typically $100-200\,\mu s$ in converter-grade thyristors).

When a thyristor is conducting it approximates to a closed switch, with a forward drop of only one or two volts over a wide range of current. Despite the low volt drop in the on state, heat is dissipated, and heatsinks must usually be provided, perhaps with fan cooling. Devices must be selected with regard to the voltages to be blocked and the r.m.s and peak currents to be carried. Their overcurrent capability is very limited, and it is usual in drives for devices to have to withstand perhaps twice full-load current for a few seconds only. Special fuses must be fitted to protect against heavy fault currents.

The reader may be wondering why we need the thyristor, since in the previous section we discussed how a transistor could be used as an electronic switch. On the face of it the transistor appears even better than the thyristor because it can be switched off while the current is flowing, whereas the thyristor will remain on until the current through it has been reduced to zero by external means. The primary reason for the use of thyristors is that they are cheaper and their voltage and current ratings extend to higher levels than power transistors. In addition, the circuit configuration in rectifiers is such that there is no need for the thyristor to be able to interrupt the flow of current, so its inability to do so is no disadvantage.

Of course there are other circuits (see for example the next section dealing with inverters) where the devices need to be able to switch off on demand, and the transistor then has the edge over the straightforward thyristor.

Single pulse rectifier

The simplest phase-controlled rectifier circuit is shown in Figure 2.6. When the supply voltage is positive, the thyristor blocks forward current until the gate pulse arrives, and up to this point the voltage across the resistive load is zero. As soon as the device turns on the voltage across it falls to near zero, and the load voltage becomes equal to the supply voltage. When the supply voltage reaches zero, so does the current. At this point the thyristor regains its blocking ability, and no current flows during the negative half-cycle.

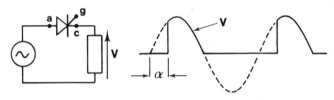

Figure 2.6 *Simple single-pulse thyristor-controlled rectifier, with resistive load and firing-angle delay a*

The load voltage (Figure 2.6) thus consists of parts of the positive half-cycles of the a.c. supply voltage. It is obviously not smooth, but is 'd.c.' in the sense that it has a positive mean value; and by varying the delay angle (a), measured from the zero crossing of the supply voltage, the mean voltage can be controlled.

The arrangement shown in Figure 2.6 gives only one peak in the rectified output for each complete cycle of the mains, and is therefore known as a 'single-pulse' or half-wave circuit. The output voltage (which ideally we would like to be steady d.c.) is so poor that this circuit is never used in drives. Instead, drive converters use four or six thyristors, and produce two or six pulse outputs, as will be seen in the following sections.

Single-phase fully-controlled converter – output voltage and control

The main elements of the converter circuit are shown in Figure 2.7. It comprises four thyristors, connected in bridge formation. (The term bridge stems from early four-arm measuring circuits, though quite why such circuits were thought to resemble a bridge seems a mystery.)

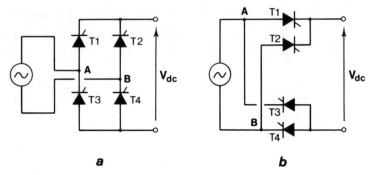

Figure 2.7   *Single-phase full-wave controlled rectifier*

The conventional way of drawing the circuit is shown in Figure 2.7(a), while in Figure 2.7(b) it has been redrawn to assist understanding. The top of the load can be connected (via T1) to terminal A of the mains, or (via T2) to terminal B of the mains, and likewise the bottom of the load can be connected either to B or to A via T3 or T4 respectively.

We are naturally interested to find what the output voltage waveform on the d.c. side will look like, and in particular to discover how it can be controlled by varying the firing delay angle *a*. This turns out to be more tricky than we might think, because the voltage waveform for a given *a* depends on the nature of the load. It is quite straightforward for resistive loads, and for loads which are sufficiently inductive for the current to remain continuous. We will therefore look first at the case where the load is resistive, and explore the basic mechanism of phase control. Later, we will see how the converter behaves with a typical motor load.

Resistive load

Thyristors T1 and T4 are fired together when terminal A of the supply is positive, while on the other half-cycle, when B is positive, thyristors T2 and T3 are fired simultaneously. The output voltage V_a, is shown by the solid line in Figure 2.8. There are two pulses per mains cycle, hence the description '2-pulse' or full-wave. At every instant the load is either connected to the mains by the pair of switches T1 and T4, or it is connected the other way up by the pair of switches T2 and T3, or it is disconnected. The load voltage therefore consists of rectified chunks of the mains voltage. It is much smoother than in the single-pulse circuit, though again it is far from pure d.c.

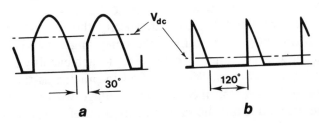

Figure 2.8 *Output voltage waveforms of single-phase fully-controlled rectifier with resistive load, for firing angle delays of 30° and 120°*

The waveform in Figure 2.8(a) corresponds to $a = 30°$, while Figure 2.8(b) is for $a = 120°$. The mean value, V_{dc}, is shown in each case. It is clear that the larger the delay angle, the lower the output voltage. The maximum output voltage (V_{do}) is obtained with $a = 0°$, and is given by

$$V_{do} = \frac{2}{\pi}\sqrt{2}V_{rms}. \qquad (2.3)$$

We note that when a is zero, the output voltage is the same as it would be for an uncontrolled diode bridge rectifier, since the thyristors conduct for the whole of the half-cycle for which they are forward-biased. The variation of the mean d.c. voltage with a is given by

$$V_{dc} = \frac{1}{2}(1 + \cos a)V_{do} \qquad (2.4)$$

from which we see that with a resistive load the d.c. voltage can be varied from a maximum of V_{do} down to zero by varying a from 0° to 180°.

Inductive (motor) load

As mentioned above, motor loads are inductive, and we have seen earlier that the current cannot change instantaneously in an inductive load. We must therefore expect the behaviour of the converter with an inductive load to differ from that with a resistive load, in which the current was seen to change instantaneously.

At first sight the fact that the behaviour of the converter depends on the nature of the load is a most unwelcome prospect. What we would like is to be able to say that, regardless of the load, we can specify the output voltage waveform once we have fixed the delay angle a. We would then know what value of a to select to achieve any desired mean output voltage. What we find in practice is that once we have fixed a, the mean output voltage with a resistive-inductive load is not the same as with a purely resistive load, and therefore we cannot give a simple general formula for the mean output voltage in terms of a.

Fortunately however, it turns out that the output voltage waveform for a given a does become independent of the load inductance once there is sufficient inductance to prevent the load current from ever falling to zero. This condition is known as 'continuous current', and, happily, most motor circuits have sufficient self-inductance to ensure that we do achieve continuous current. Under continuous current conditions, the output voltage waveform only depends on the firing angle, and not on the actual inductance present. This makes things much more straightforward, and typical output voltage waveforms for this continuous current condition are shown in Figure 2.9.

The waveform in Figure 2.9(a) corresponds to $a = 15°$, while Figure 2.9(b) corresponds to $a = 60°$. We see that, as with the resistive load, the larger the delay angle the lower the mean output voltage. However with the resistive load the output voltage was never negative, whereas we see that for short periods the output voltage can now become negative. This is because the inductance smooths out the current so

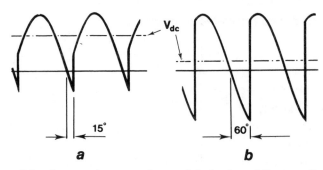

Figure 2.9   *Output voltage waveforms of single-phase fully-controlled rectifier supplying an inductive (motor) load, for firing angle delays of 15° and 60°*

that at no time does it fall to zero. As a result, one or other pair of thyristors is always conducting, and there is no time at which the load is disconnected from the a.c. supply. We will see in Chapter 4 that although the output voltage is far from pure d.c., the d.c. motor can operate perfectly happily when supplied from a thyristor converter, its speed being determined by the average output voltage.

The maximum voltage (V_{do}) is again obtained when a is zero, and is the same as for the resistive load (equation 2.3). It is easy to show that the mean d.c. voltage is now related to a by

$$V_{dc} = V_{do}\cos a \qquad (2.5)$$

This equation indicates that we can control the mean output voltage by controlling a. We also see that when a is greater than 90° the mean output voltage is negative. The fact that we can obtain a nett negative output voltage with an inductive load contrasts sharply with the resistive load case, where the output voltage could never be negative. We shall see later that this facility allows the converter to return energy from the load to the supply, and this is important when we want to use the converter with a d.c. motor in the regenerating mode.

3-phase fully-controlled converter

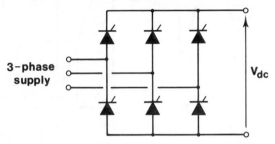

Figure 2.10   *Three-phase fully-controlled thyristor converter*

The main power elements are shown in Figure 2.10. The three-phase bridge has only two more thyristors than the single-phase, but the output voltage waveform is very much better, as shown in Figure 2.11. There are now six pulses of the output voltage per mains cycle, hence the description '6-pulse'. The thyristors are again fired in pairs (one in the top half of the bridge and one in the bottom half), and each thyristor carries the output current for one third of the time. As in the single-phase converter, the delay angle controls the output voltage, but now $a = 0$ corresponds to the point at which the phase voltages are equal.

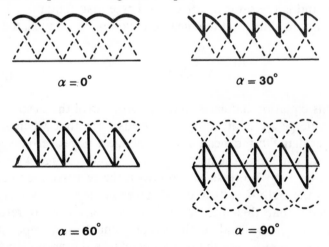

Figure 2.11   *Output voltage waveforms for three-phase fully-controlled converter supplying an inductive (motor) load*

The enormous improvement in the smoothness of the output voltage waveform is clear when we compare Figures 2.11 and 2.9, and it indicates the wisdom of choosing a 3-phase converter whenever possible. The much better voltage waveform also means that the desirable 'continuous current' condition is much more likely to be met, and the waveforms in Figure 2.11 have therefore been drawn with the assumption that the load current is in fact continuous. Occasionally, even a six-pulse waveform is not sufficiently smooth, and some very large drive converters therefore consist of two six-pulse converters with their outputs in series. A phase-shifting transformer is used, to insert a 30° shift between the a.c. supplies to the two 3-phase bridges. The resultant ripple voltage is then 12-pulse.

Returning to the 6-pulse converter, the mean output voltage can be shown to be given by

$$V_{dc} = V_{do}\cos a = \frac{3}{\pi} \sqrt{2}V_{rms}\cos a. \qquad (2.6)$$

We note that we can obtain the full range of output voltages from $+V_{do}$ to $-V_{do}$, so as with the single-phase converter, regenerative operation will be possible.

Output voltage range

In Chapter 4 we will discuss the use of the fully-controlled converter to drive a d.c. motor, so it is appropriate at this stage to look briefly at the typical voltages we can expect. Mains a.c. supply voltages obviously vary around the world, but single-phase supplies are usually 220–240 V, and we see from equation 2.3 that the maximum mean d.c. voltage available from a single phase 240 V supply is 216 V. This is suitable for 180–200 V motors. If a higher voltage is needed (for say a 300 V motor), a transformer must be used to step up the mains.

Turning now to typical three-phase supplies, the lowest three-phase industrial voltages are usually around 380–440 V. (Higher voltages of up to 11 kV are used for large

drives, but these will not be discussed here). So with V_{rms} =
415 V for example, the maximum d.c. output voltage (equation 2.6) is 560 volts. After allowances have been made for
supply variations and impedance drops, we could not rely
on obtaining much more than 520–540 V, and it is usual for
the motors used with 6-pulse drives fed from 415 V, 3-phase
supplies to be rated in the range 440–500 V. (Often the field
winding will be supplied from single phase 240 V, and field
voltage ratings are then around 180–200 V, to allow a mar-
gin in hand from the theoretical maximum of 216 V referred
to earlier.)

Firing circuits

Since the gate pulses are only of low power, the gate drive
circuitry is simple and cheap. Often a single integrated circuit
(chip) contains all the circuitry for generating the gate pulses, and for synchronizing them with the appropriate delay
angle a with respect to the applied voltage. To avoid direct
electrical connection between the high voltages in the main
power circuit and the low voltages used in the control
circuits, the gate pulses are usually coupled to the thyristor
by means of small pulse transformers. Most converters include inverse cosine-weighted firing circuitry, and the output
voltage then varies in direct proportion to an input control
voltage of say 0–10 volts.

A.C. FROM D.C. – INVERSION

The business of getting a.c. from d.c. is known as inversion,
and nine times out of ten we would ideally like to be able to
produce sinusoidal voltages of whatever frequency and
amplitude we choose. Unfortunately the constraints imposed
by the necessity to use a switching strategy mean that we
always have to settle for a voltage waveform which is
composed of rectangular chunks, and is thus far from ideal.
Nevertheless it turns out that a.c. motors are remarkably
tolerant, and are happy to operate despite the inferior waveforms produced by the inverter.

Single-phase inverter

We can illustrate the basis of inverter operation by considering the single-phase example shown in Figure 2.12.

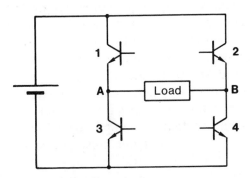

Figure 2.12   *Inverter circuit for single-phase output*

The input or d.c. side of the inverter is usually referred to as the 'd.c. link', reflecting the fact that in the majority of cases the d.c. is obtained by rectifying and smoothing the incoming constant-frequency mains. The output or a.c. side is taken from terminals A and B in Figure 2.12.

When transistors 1 and 4 are switched on, the load voltage is positive, and equal to the d.c. link voltage, while when 2 and 3 are on it is negative. If no devices are switched on, the output voltage is zero. Typical output voltage waveforms at low and high frequencies are shown in Figures 2.13(a) and 2.13(b) respectively.

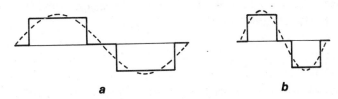

Figure 2.13   *Inverter output waveforms – resistive load*

Here each pair of devices is on for one-third of a cycle, and all the devices are off for two periods of one-sixth of a cycle. The output waveform is clearly not a sinewave, but at least it is alternating and symmetrical. The fundamental component is shown dotted in Figure 2.13.

Within each cycle the pattern of switching is regular, and easily programmmed using appropriate logic circuitry. Frequency variation is obtained by altering the clock frequency controlling the 4-step switching pattern. (The oscillator providing the clock signal can be controlled by an analogue voltage, or it can be generated in software.) The effect of varying the switching frequency is shown in Figure 2.13, from which we can see that the amplitude of the fundamental component of voltage remains constant, regardless of frequency. Unfortunately (as explained in Chapter 7) this is not what we want for supplying an induction motor. To prevent the flux from falling as the frequency is raised we need to be able to increase the voltage in proportion to the frequency. We will look at voltage control shortly, after a brief digression to discuss the problem of 'shoot-through'.

Inverters with the configurations shown in Figures 2.12 and 2.15 are subject to a potentially damaging condition which can arise if both transistors in one 'leg' of the inverter inadvertently turn 'on' simultaneously. This should never happen if the devices are switched correctly, but if something goes wrong and both devices are on together – even for a very short time – they form a short-circuit across the d.c. link. This fault condition is referred to as 'shoot-through' because a high current is established very rapidly, destroying the devices. A good inverter therefore includes provision for protecting against the possibility of shoot-through, usually by imposing a minimum time-delay between one device in the leg going off and the other coming on.

Output voltage control

There are two ways in which the voltage can be controlled. First, if the d.c. link is provided from a.c. mains via a controlled rectifier or from a battery via a chopper, the d.c. link voltage can be varied. We can then set the amplitude of the output voltage to any value within the range of the link. In particular, we can arrange for the link voltage to track the output frequency of the inverter, so that at high output frequency we obtain a high output voltage and vice-versa. This method of voltage control results in a simple inverter, but requires a controlled (and thus relatively expensive) rectifier for the d.c. link.

The second method, which now predominates in small and medium sizes, achieves voltage control by pulse-width-modulation (PWM) within the inverter itself. A cheaper uncontrolled rectifier can thus be used to provide a constant-voltage d.c. link.

The principle of voltage control by PWM is illustrated in Figure 2.14.

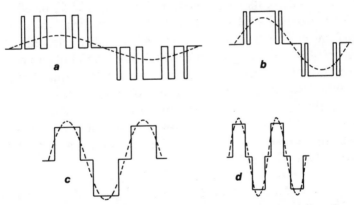

Figure 2.14   *Inverter output voltage and frequency control by pulse-width modulation*

At low output frequencies, a low output voltage is required, so one of each pair of devices is used to chop the voltage, the mark-space ratio being varied to achieve the desired voltage at the output. The low fundamental voltage component at low frequency is shown dotted in Figure

2.14(a). At a higher frequency a higher voltage is needed, so the chopping device is allowed to conduct for a longer period, giving the higher fundamental output shown in Figure 2.14(b). As the frequency is raised still higher, the separate 'on' periods eventually merge, giving the waveform shown in Figure 2.14(c). Any further increase in frequency takes place without further increase in the output voltage, as shown in Figure 2.14(d).

In drive applications, the range of frequencies over which the voltage/ frequency ratio can be kept constant is known as the 'PWM' region, and the upper limit of the range is usually taken to define the 'base speed' of the motor. Above this frequency, the inverter can no longer match voltage to frequency, the inverter effectively having run out of steam as far as voltage is concerned. The maximum voltage is thus governed by the link voltage, which must therefore be sufficiently high to provide whatever fundamental voltage the motor needs at its base speed, which is usually 50 or 60 Hz.

Beyond the PWM region the voltage waveform is as shown in Figure 2.14(d): this waveform is invariably referred to as 'quasi-square', though in the light of the overall object of the exercise a more accurate description would be 'quasi-sine'.

When supplying an inductive motor load, fast recovery freewheel diodes are needed in parallel with each device. These may be discrete devices, or fitted in a common package with the transistor, or even integrated to form a single transistor/diode device.

Sinusoidal PWM

So far we have emphasized the importance of being able to control the amplitude of the fundamental output voltage by modulating the width of the pulses which make up the output waveform. If this was the only requirement, we would have an infinite range of modulation patterns which would do. But as well as the right fundamental amplitude, we want the harmonic content to be minimized. It is particularly important to limit the amplitude of the low-order harmonics, since these are the ones which the motor is most likely to respond to.

The number, width, and spacing of the pulses is therefore optimized to keep the harmonic content as low as possible. A host of sophisticated strategies have been developed, almost all using a microprocessor based system to store and/or generate the modulation patterns. There is an obvious advantage in using a high switching frequency, since there are then more pulses to play with. Ultrasonic frequencies are already in use, and as devices improve the switching frequencies will go still higher. Most manufacturers claim their system is better than the competition, but it is too early to say which will ultimately prove 'best' for motor operation. Some early schemes used comparatively few pulses per cycle, and changed the number of pulses in discrete steps rather than smoothly, which earned them the nickname 'gear-changers'. These inverters were noisy and irritating.

3-phase inverter

A 3-phase output can be obtained by adding only two more switches to the four needed for a single-phase inverter, giving the typical power-circuit configuration shown in Figure 2.15.

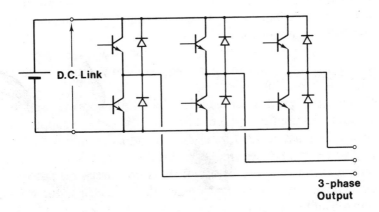

Figure 2.15   *Three-phase inverter power circuit*

A freewheel diode is required in parallel with each transistor to protect against overvoltages caused by an inductive (motor) load.

We note that the circuit configuration in Figure 2.15 is the same as as for the 3-phase controlled rectifier looked at earlier. We mentioned then that the controlled rectifier could be used to regenerate, i.e. to convert power from d.c. to a.c., and this is of course 'inversion' as we now understand it.

A variety of switching devices are currently used in variable-frequency inverters for induction motor drives, including bipolar transistors, gate-turn-off thyristors (GTOs), and various types of field-effect transistors, as well as hybrid (bipolar/field-effect-IGBT) types. With new switching devices continually being introduced, and with the move to integrate all the base or gate drive and protection circuitry within the main device package (the so-called 'Smart-Power' approach), it will be some time before an industry standard emerges.

Forced and natural commutation

We have assumed in this discussion that the switching devices can turn-off or 'commutate' on demand, so that the output (load) current is either reduced to zero or directed to another leg of the inverter. Transistors and GTOs can operate like this, but, as explained earlier, conventional thyristors cannot turn off on command. Nevertheless thyristors are widely used to invert power from d.c. to a.c. as we will see when we look at the d.c. motor drive in Chapter 4, so we must be clear how this is possible.

There are two distinct ways in which thyristors are used in inverters. In the first, where the inverter is used to supply an essentially passive load, such as an induction motor, each thyristor has to be equipped with its own auxiliary 'forced commutating' circuit, whose task is to force the current through the thyristor to zero when the time comes for it to turn off. The commutating circuits are quite complex, and require substantial capacitors to store the energy required for commutation. Force commutated thyristor converters therefore tend to be bulky and expensive. At one time they were the only alternative, but are now rapidly being superseded by transistor or GTO versions.

The other way in which thyristors can be used to invert power from d.c. to a.c. is when the a.c. side of the bridge is connected to a 3-phase mains supply. This is the normal 'controlled rectifier' arrangement introduced earlier. In this case it turns out that the currents in the thyristors are naturally forced to zero by the active mains voltages, thereby allowing the thyristors to commutate or turn off naturally. This mode of operation continues to be important, as we will see when we look at d.c. motor drives.

CONVERTER WAVEFORMS AND ACOUSTIC NOISE

In common with most textbooks, the waveforms shown in this chapter (and later in the book) are what we would hope to see under ideal conditions. It makes sense to concentrate on these ideal waveforms from the point of view of gaining a basic understanding, but we ought to be warned that what we see on an oscilloscope may well look rather different!

We have seen that the essence of power electronics is the switching process, so it should not come as much of a surprise to learn that in practice the switching is seldom achieved in such a clear-cut fashion as we have assumed. Usually, there will be some sort of high-frequency 'ringing' evident, particularly on the voltage waveforms following each transition due to switching. This is due to the effects of stray capacitance and inductance: it will have been anticipated at the design stage, and steps will have been taken to minimise it by fitting 'snubbing' circuits at the appropriate places in the converter. However complete suppression of all these transient phenomena is seldom economically worthwhile so the user should not be too alarmed to see remnants of the transient phenomena in the output waveforms.

Acoustic noise is also a matter which can worry newcomers. Most power electronic converters emit whining or humming sounds at frequencies corresponding to the fundamental and harmonics of the switching frequency, though when the converter is used to feed a motor, the sound of the motor is usually a good deal greater than the sound from the converter itself. These sounds are very difficult to describe in words, but typically range from a high-pitched hum through a whine to a piercing whistle. They vary in intensity with the size of converter and the load, and to the trained ear can give a good indication of the health of the motor and converter.

COOLING OF POWER SWITCHING DEVICES

Thermal resistance

We have seen that by adopting a switching strategy the power loss in the switching devices is small in comparison with the power throughput, so the converter has a high efficiency. Nevertheless almost all the heat which is produced in the switching devices is released in the active region of the semiconductor, which is itself very small and will overheat and breakdown unless it is adequately cooled. It is therefore important to ensure that even under the most onerous operating conditions, the temperature of the active junction inside the device does not exceed the safe value.

Consider what happens to the temperature of the junction region of the device when we start from cold (i.e. ambient) temperature and operate the device so that its average power dissipation remains constant. At first, the junction temperature begins to rise, so some of the heat generated is conducted to the metal case, which stores some heat as its temperature rises. Heat then flows into the heatsink (if fitted) which begins to warm up, and heat begins to flow to the surrounding air, at ambient temperature. The temperatures of the junction, case and heatsink continue to rise until eventually an equilibrium is reached when the total rate of loss of heat to ambient temperature is equal to the power dissipation inside the device.

The final steady-state junction temperature thus depends on how difficult it is for the power loss to escape down the temperature gradient to ambient, or in other words on the total 'thermal resistance' between the junction inside the device and the surrounding medium (usually air). Thermal resistance is usually expressed in °C/watt, which directly indicates how much temperature rise will occur in the steady state for every watt of heat flow. It follows that for a given power dissipation, the higher the thermal resistance, the higher the temperature rise, so in order to minimize the temperature rise of the device, the total thermal resistance between it and the surrounding air must be made as small as possible.

The device designer aims to minimize the thermal resistance between the semiconductor junction and the case of the device, and provides a large and flat metal mounting surface to minimize the thermal resistance between the case and the heatsink. The converter designer must ensure good thermal contact between the device and the heatsink, usually by a bolted joint liberally smeared with heat-conducting compound to fill any microscopic voids, and must design the heatsink to minimize the thermal resistance to air (or in some cases oil or water). Heatsink design offers the only real scope for appreciably reducing the total resistance, and involves careful selection of the material, size, shape and orientation of the heatsink, and the associated air-moving system (see below).

One drawback of the good thermal path between the junction and case of the device is that the metal mounting surface (or surfaces in the case of the popular hockey-puck package) are electrically live. This poses a difficulty for the converter designer, because mounting the device directly on the heatsink causes the latter to be electrically live and therefore hazardous. In addition, several separate isolated heatsinks may be required in order to avoid short-circuits. The alternative is for the devices to be electrically isolated from the heatsink using thin mica spacers, but then the thermal resistance is appreciably increased.

The increasing trend towards packaged modules with an electrically isolated metal base gets round this problem. These contain combinations of transistors, diodes or thyristors, from which various converter circuits can be built up. Several modules can be mounted on a single heatsink, which does not have to be isolated from the enclosure or cabinet. Currently they are available in ratings suitable for converters up to many tens of kW, and the range is expanding. This development, coupled with a move to fan-assisted cooling of heatsinks has resulted in a dramatic reduction in the overall size of complete converters, so that a modern 20 kW drive converter is perhaps only the size of a small suitcase.

Arrangement of heatsinks and forced-air cooling

The principal factors which govern the thermal resistance of a heatsink are the total surface area, the condition of the surface and the air flow. Most converters use extruded aluminium heatsinks, with multiple fins to increase the effective cooling surface area and lower the resistance, and with a machined face or faces for mounting the devices. Heatsinks are usually mounted vertically to improve natural air convection. Surface finish is important, with black anodised aluminium being typically 30 per cent better than bright. At one time heatsinks could only be bought from the manufacturer, but a wide range of compact and highly efficient heatsinks and fans designed specifically for cooling electronic equipment is now readily available from the leading electronic component suppliers.

A typical layout for a medium-power (say 200 kW) converter is shown in Figure 2.16.

The fan(s) are positioned either at the top or bottom of the heatsink, and draw external air upwards, assisting natural convection. The value of even a modest air-flow is shown by the sketch in Figure 2.17. With a velocity of only 2 m/sec, for example, the thermal resistance is halved as compared with

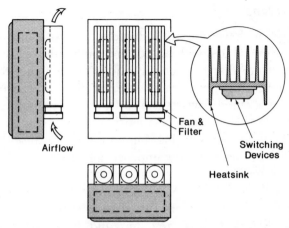

Figure 2.16 *Typical layout of converter showing heatsinks and cooling fans*

the naturally-cooled set-up, which means that for a given temperature rise the heatsink can be half the size of the naturally-cooled one. Only a little of this saving in space is taken up by the fans, as shown Figure 2.16. Large increases in the air velocity bring diminishing returns, as shown in Figure 2.17, and also introduce additional noise which is generally undesirable.

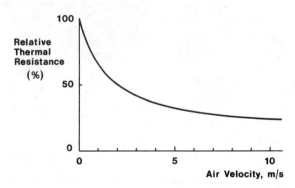

Figure 2.17 *Sketch indicating influence of air velocity on effective thermal resistance. (The thermal resistance in still air is taken as 100 per cent.)*

Cooling fans

Cooling fans have integral hub-mounted inside-out motors, i.e. the rotor is outside the stator and carries the blades of the fan. The rotor diameter/length ratio is much higher than for most conventional motors in order to give a slimline shape to the fan assembly, which is well-suited for mounting at the end of an extruded heatsink (Figure 2.16). The rotor inertia is thus relatively high, but this is unimportant because the total inertia is dominated by the impeller, and there is no need for high accelerations.

Mains voltage 50 or 60 Hz fans have external-rotor single-phase shaded-pole motors, which normally run at a fixed speed of around 2700 rev/min, and have input powers typically between 10 and 50 W. The torque required in a fan

Plate 2.1   *Range of d.c. and a.c. motor drive converters ranging from 500 W to 50 kW. The low-power control circuitry is visible at the front, the power switching devices being mounted on heatsinks at the rear. The heatsinks and cooling fans are visible on the larger units (**Photograph by courtesy of GEC Industrial Controls Ltd**)*

is roughly proportional to the cube of the speed, so the starting torque requirement is low and the motor can be designed to have a high running efficiency. (See Chapter 6.) Where acoustic noise is a problem slower-speed (but less efficient) versions are used.

Low-voltage (5, 12 or 24 V) d.c. fans employ brushless motors with Hall-effect rotor position detection (Chapter 9). The absence of sparking from a conventional commutator is important to limit interference with adjacent sensitive equipment. These fans are generally of lower power than their a.c. counterparts, typically from as little as 1 W up to about 10 W, and with running speeds of typically between 3000 and 5000 rev/min. They are mainly used for cooling circuit boards directly, and have the advantage that the speed can be controlled by voltage variation, thereby permitting a trade-off between noise and volume flow.

3

CONVENTIONAL D.C. MOTORS

INTRODUCTION

Until the 1980s the conventional (brushed) d.c. machine was the automatic choice where speed or torque control is called for, and it remains pre-eminent in spite of the growing challenge from the inverter-fed induction motor. Applications range from steel rolling mills and railway traction, through a very wide range of industrial drives through to robotics, printers and precision servos. The range of power outputs is correspondingly wide, from several megawatts at the top end down to a only a few watts, but except for a few of the small low-performance ones, such as those used in toys, all have the same basic structure, as shown in Figure 3.1.

As in any electrical machine it is possible to design a d.c. motor for any desired supply voltage, but for several reasons it is unusual to find rated voltages lower than about 6 V or much higher than 700 V. The lower limit arises because the brushes (see below) give rise to an unavoidable volt-drop of perhaps 0.5–1 V, and it is clearly not good practice to let this 'wasted' voltage became a large fraction of the supply voltage. At the other end of the scale it becomes prohibitively expensive to insulate the commutator segments to withstand higher voltages. The function and operation of the commutator is discussed later, but it is appropriate to

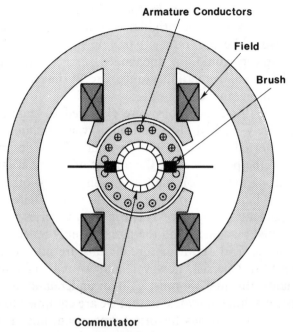

Figure 3.1   *Conventional (brushed) d.c. motor*

mention here that brushes and commutators are troublesome at very high speeds. Small d.c. motors, say up to a hundreds of watts output, can run at perhaps 12 000 rev/min, but the majority of medium and large motors are usually designed for speeds below 3 000 rev/min.

Increasingly, motors are being supplied with power-electronic drives, which draw power from the a.c. mains and convert it to d.c. for the motor. Since the mains voltages tend to be standardized (e.g. 110 V, 220–240 V, or 380–440 V, 50 or 60 Hz), motors are made with rated voltages which match the range of d.c. outputs from the converter (see Chapter 2).

As mentioned above, it is quite normal for a motor of a given power, speed and size to be available in a range of different voltages. In principle all that has to be done is to alter the number of turns and the size of wire making up the coils in the machine. A 12 V, 4 A motor, for example, could easily be made to operate from 24 V instead, by winding its coils with twice as many turns of wire having only half the cross-sectional area of the original. The full speed would be the same at 24 V as the original was at 12 V, and the rated current would be 2 A, rather than 4 A. The input power and output power would be unchanged, and externally there would be no change in appearance, except that the terminals might be a bit smaller.

Traditionally d.c. motors were classified as shunt, series, or separately excited. In addition it was common to see motors referred to as 'compound-wound.' These descriptions date from the period before the advent of power electronics, and a strong association built up linking one or other 'type' of d.c. machine with a particular application. There is really no fundamental difference between shunt, series or separately excited machines, and the names simply reflect the way in which the field and armature circuits are interconnected. The terms still persist, however, and we will refer to them again later. But first we must gain an understanding of how the basic machine operates, so that we are equipped to understand what the various historic terms mean, and hence see how modern practice is deployed to achieve the same ends.

We should make clear at this point that whereas in an a.c. machine the number of poles is of prime importance in determining the speed, the pole-number in a d.c. machine is of little consequence as far as the user is concerned. It turns out to be more economical to use two or four poles in small or medium size d.c. motors, and more (e.g. ten or twelve or even more) in large ones, but the only difference to the user is that the 2-pole type will have 2 brushes at 180°, the 4-pole will have 4 brushes at 90°, and so on. Most of our discussion centres on the 2-pole version in the interests of simplicity, but there is no essential difference as far as operating charactersitics are concerned.

TORQUE PRODUCTION

Torque is produced by interaction between the axial current-carrying conductors on the rotor and the radial magnetic flux

Plate 3.1 *Permanent-magnet d.c. motor, 500 W at 2000 rev/min*
(*Photograph by courtesy of GEC Electromotors Ltd*)

produced by the stator. This flux or 'excitation' can be fur-
nished by permanent magnets (Figure 3.2(a)) or by means of
field windings (Figure 3.2(b)).

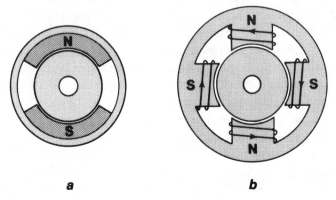

a   **b**

Figure 3.2 *Excitation (field) systems for d.c. motors (a) 2-pole permanent
magnet; (b) 4-pole wound field*

Permanent magnet machines are available in motors with outputs from a few watts up to a few kilowatts, while wound-field machines begin at about 100 watts and extend to the largest outputs. The advantages of the permanent magnet type are that no electrical supply is required for the field, and the overall size of the motor can be smaller. On the other hand the strength of the field cannot be varied so one possible option for control is ruled out.

Ferrite magnets have been used for many years. They are relatively cheap and easy to manufacture but their energy product (a measure of their effectiveness as a source of excitation) is poor. Rare earth magnets (e.g. Samarium Cobalt) provide much higher energy products, and open up the possibility of very high torque/volume ratios. They are used in high-performance servo motors, but are relatively expensive, difficult to manufacture and handle, and the raw material is found in politically sensitive areas. The latest 'super' magnet material is Neodymium Iron Boron, which has the highest energy product and is predicted to revolutionize motor design. At present its major handicap is that it can only be operated at temperatures below about 150°, which is not sufficient for some high-performance motors.

Although the magnetic field is essential to the operation of the motor, we should recall that in Chapter 1 we saw that none of the mechanical output power actually comes from the field system. The excitation acts rather like a catalyst in a chemical reaction, making the energy conversion possible but not contributing to the output.

The main (power) circuit consists of a set of identical coils wound in slots on the rotor, and known as the armature. Current is fed into and out of the rotor via carbon brushes which make sliding contact with the 'commutator', which consists of insulated copper segments mounted on a cylindrical former.

The function of the commutator is discussed further below, but it is worth stressing here that all the electrical energy which is to be converted into mechanical output has to be fed into the motor through the brushes and commutator. Given that a high-speed sliding contact is involved, it is not surprising that to ensure trouble-free operation the commutator needs to be kept clean, and the brushes and their associated springs need to be regularly serviced. Brushes wear away, of course, though if properly set they can last for thousands of hours. All being well the brush debris (in the form of graphite particles) will be carried out of harm's way by the ventilating air: any build-up of dust on a greasy commutator is dangerous and can lead to disastrous 'flashover' faults.

The axial length of the commutator depends on the current it has to handle. Small motors usually have one brush on each side of the commutator, so the commutator is quite short, but larger heavy-current motors may well have many brushes mounted on a common arm, each with its own brushbox (in which it is free to slide) and with all the brushes on one arm connected in parallel via their flexible copper leads or 'pigtails'. The length of the commutator can then be comparable with the 'active' length of the armature.

Function of the commutator

Many different winding arrangements are used for d.c. armatures, and it is neither helpful or necessary for us to delve into the nitty-gritty of winding and commutator design. These are matters which are best left to motor designers and repairers. What we need to do is to focus on what a well-designed commutator-winding actually achieves, and despite the apparent complexity, this can be stated quite simply.

The purpose of the commutator is to ensure that regardless of the position of the rotor, the pattern of current flow in the rotor is always as shown in Figure 3.3.

Current enters the rotor via one brush, flows through all the rotor coils in the directions shown in Figure 3.3, and leaves via the other brush. The first point of contact with the armature is via the segment or segments on which the brush is pressing at the time (the brush is usually wider than a single segment), but since the interconnections between the in-

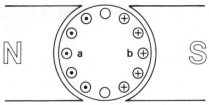

Figure 3.3   *Pattern of rotor (armature) currents in 2-pole d.c. motor*

dividual coils are made at each commutator segment, the current actually passes through all the coils via all the commutator segments in its path through the armature.

We can see from Figure 3.3 that all the conductors lying under the N pole carry current in one direction, while all those under the S pole carry current in the opposite direction. All the conductors under the N pole will therefore experience an upward force (which is proportional to the flux density B and the armature current I) while all the conductors under the S pole will experience an equal downward force. A torque is thus produced on the rotor, the magnitude of the torque being proportional to the flux density B and the current I. In practice the flux density B will not be completely uniform under the pole, so the force on some of the armature conductors will be greater than on others. However, it is straightforward to show that the total torque developed is given by

$$T = K_T \phi I \qquad (3.1)$$

where ϕ is the total flux produced by the field, and K_T is constant. In permanent magnet motors, the flux is sensibly constant, so we see that the motor torque is directly proportional to the armature current. This extremely simple result means that if a motor is required to produce constant torque at all speeds, we simply have to arrange to keep the armature current constant. Standard drive packages usually include provision for doing this, as will be seen later. We can also see from equation 3.1 that the direction of the torque can be reversed by reversing either the armature current (I) or the flux (ϕ). We obviously make use of this when we want the motor to run in reverse, and sometimes when we want regenerative braking.

The alert reader might rightly challenge the claim made above that the torque will be constant regardless of rotor position. Looking at Figure 3.3, it should be clear that if the rotor turned just a few degrees, one of the five conductors shown as being under the pole will move out into the region where there is no radial flux, before the next one moves under the pole. Instead of five conductors producing force, there will then be only four, so won't the torque be reduced accordingly?

The answer to this question is yes, and it is to limit this unwelcome variation of torque that most motors have many more coils than are shown in Figure 3.3. Smooth torque is of course desirable in most applications in order to avoid vibrations and resonances in the transmission and load, and is essential in machine tool drives where the quality of finish can be marred by uneven cutting if the torque and speed are not steady.

Broadly speaking the higher the number of coils (and commutator segments) the better, because the ideal armature would be one in which the pattern of current on the rotor corresponded to a 'current sheet', rather than a series of discrete packets of current. If the number of coils was infinite, the rotor would look identical at every position, and the torque would therefore be absolutely smooth. Obviously this is not practicable, but it is closely approximated in most d.c. motors. For practical and economic reasons the number of slots is higher in large motors, which may well have a hundred or more coils and hence very little ripple in their output torque.

Operation of the commutator – interpoles

Returning now to the operation of the commutator, and focusing on a particular coil (e.g. the one shown as ab in Figure 3.3) we note that for half a revolution – while side a is under the N pole and side b is under the S pole – the current needs to be positive in side a and negative in side b in order to produce a positive torque. For the other half revolution, while side a is under the S pole and side b is under the N pole, the current must flow in the opposite direction through the coil for it to continue to produce positive torque. This reversal of current takes place in each coil as it passes through the interpolar axis, the coil being 'switched-round' by the action of the commutator sliding under the brush.

Each time a coil reaches this position it is said to be undergoing commutation, and the relevant coil in Figure 3.3 has therefore been shown as having no current to indicate that its current is in the process of changing from positive to negative.

The essence of the current-reversal mechanism is revealed by the simplified sketch shown in Figure 3.4.

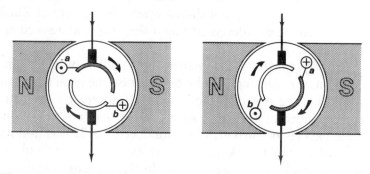

Figure 3.4   *Simplified diagram of single-coil motor to illustrate the current-reversing function of the commutator*

This diagram shows only a single coil, and it should be stressed again that in an actual multi-coil armature, only one coil is reversed at a time, and that the commutator arc is much smaller than that shown in Figure 3.4.

The main difficulty in achieving good commutation arises because of the self inductance of the armature coils, and the associated stored energy. As we saw earlier, inductive circuits tend to resist change of current, and if the current reversal has not been fully completed by the time the brush slides off the commutator segment in question there will be a spark at the trailing edge of the brush.

In small motors some sparking is considered tolerable, but in medium and large wound-field motors small additional stator poles known as interpoles (or compoles) are provided to improve commutation and hence minimize sparking. These extra poles are located midway between the main field poles, as shown in Figure 3.5. Interpoles are not normally required in permanent magnet motors because the absence of stator iron close to the rotor coils results in much lower armature coil inductance.

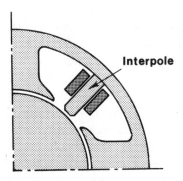

Figure 3.5 *Sketch showing location of interpole and interpole winding (the main field windings have been omitted for the sake of clarity)*

The purpose of the interpoles is to induce a motional e.m.f. in the coil undergoing commutation, in such a direction as to speed-up the desired reversal of current, and thereby prevent sparking. The e.m.f. needed is proportional to the current which has to be commutated, i.e. the armature current, and to the speed of rotation. The correct e.m.f. is therefore achieved by passing the armature current through the coils on the interpoles, thereby making the flux from the interpoles proportional to the armature current. The interpole coils therefore consist of a few turns of thick conductor, connected permanently in series with the armature.

Motional e.m.f.

When the armature is stationary, no motional e.m.f. is induced in it. But when the rotor turns, the armature conductors cut the magnetic flux and an e.m.f. is induced in them.

As far as each individual coil on the armature is concerned, an alternating e.m.f. will be induced in it when the rotor rotates. For the coil ab in Figure 3.3, for example, side a will be moving upward through the flux if the rotation is clockwise, and an e.m.f. directed out of the plane of the paper will be generated. At the same time the return side of the coil (b) will be moving downwards, so the same magnitude of e.m.f. will be generated, but directed into the paper. The resultant e.m.f. in the coil will therefore be twice that in the coil-side, and this e.m.f. will remain constant for almost half a revolution, during which time the coil sides are cutting a constant flux density. For the comparatively short time when the coil is not cutting any flux the e.m.f. will be zero, and then the coil will begin to cut through the flux again, but now each side is under the other pole, so the e.m.f. is in the opposite direction. The resultant e.m.f. waveform in each coil is therefore a rectangular alternating wave, with magnitude and frequency proportional to the speed of rotation.

The coils on the rotor are connected in series, so if we were to look at the e.m.f. across any given pair of diametrically opposite commutator segments, we would see a large alternating e.m.f. (We would have to station ourselves on the rotor to do this, or else make sliding contacts using slip-rings.)

The fact that the induced voltage in the rotor is alternating may come as a surprise, since we are talking about a d.c. motor rather than an a.c. one. But any worries we may have should be dispelled when we ask what we will see by way of induced e.m.f. when we 'look in' at the brushes. We shall see that the brushes and commutator effect a remarkable transformation, bringing us back into the reassuring world of d.c.

The first point to note is that the brushes are stationary. This means that although a particular segment under each brush is continually being replaced by its neighbour, the circuit lying between the two brushes always consists of the same number of coils, with the same orientation with respect to the poles. As a result the e.m.f. at the brushes is direct, rather than alternating.

The magnitude of the e.m.f. depends on the position of the brushes around the commutator, but they are invariably placed at the point where they continually 'see' the peak value of the alternating e.m.f. induced in the armature. In effect, the commutator and brushes can be regarded as a mechanical rectifier which converts the alternating e.m.f. in the rotating reference frame to a direct e.m.f. in the stationary reference frame. It is a remarkably clever and effective device, its only real drawback being that it is a mechanical system, and therefore subject to wear and tear.

We saw earlier that to obtain smooth torque it was necessary for there to be a large number of coils and commutator segments, and we find that much the same considerations apply to the smoothness of the generated e.m.f. If there are only a few armature coils the e.m.f. will have a noticeable ripple superimposed on the mean d.c. level. The higher we make the number of coils, the smaller the ripple, and the better the d.c. we produce. The small ripple we inevitably get with a finite number of segments is seldom any problem with motors used in drives, but can give rise to difficulties when a d.c. machine is used as a tachogenerator.

From the discussion of motional e.m.f. in Chapter 1, it follows that the magnitude of the resultant e.m.f. (E) which is generated at the brushes is proportional to the flux (ϕ) and the speed (n), and is given by

$$E = K_E \phi n \qquad (3.2)$$

where K_E is constant for the motor in question.

This equation reminds us of the key role of the flux, in that until we switch on the field no voltage will be generated, no matter how fast the rotor turns. Once the field is energized, the generated voltage is directly proportional to the speed of rotation, so if we reverse the direction of rotation, we will also reverse the polarity of the generated e.m.f. We should also remember that the e.m.f. depends only on the flux and the speed, and is the same regardless of whether the rotation is provided by some external source (i.e. when the machine is being driven as a generator) or when the rotation is produced by the machine itself (i.e. when it is acting as a motor).

When the flux ϕ is at its full value, equations (3.1) and (3.2) can be written in the form

$$T = k_t I \qquad (3.3)$$

$$E = k_e \omega \qquad (3.4)$$

where k_t is the motor torque constant, k_e is the e.m.f. constant, and ω is the angular speed in rad/sec. The S.I. units for k_t are Nm/A, and for k_e the units are volts/rad/sec. (Note, however, that k_e is more often given in volts/1000 rev/min.) In S.I. units, the torque and e.m.f. constants are equal, i.e. $k_t = k_e = k$. The torque and e.m.f. equations can thus be written

$$T = kI \qquad (3.5)$$

$$E = kn. \qquad (3.6)$$

Equivalent circuit

The equivalent circuit can now be drawn on the same basis as we used for the primitive machine in Chapter 1, and is shown in Figure 3.6.

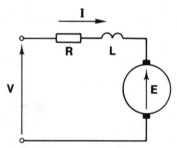

Figure 3.6 *Equivalent circuit of a d.c. motor*

The sign convention adopted in Figure 3.6 is the usual one when the machine is operating as a motor. The voltage V is the voltage applied to the armature terminals (i.e. across the brushes), and E is the internally developed motional e.m.f. Under motoring conditions, E always opposes the applied voltage V, and for this reason it is referred to as back e.m.f. For current to be forced into the motor, V must be greater than E, the voltage equation being given by

$$V = E + IR + L di/dt \qquad (3.7)$$

where R is the resistance of the armature and L is the self inductance.

D.C. MOTOR − STEADY-STATE CHARACTERISTICS

From the user's viewpoint the extent to which speed falls when load is applied, and the variation in speed with applied voltage are usually the first questions which need to be answered in order to assess the suitability of the motor for the job in hand. The information is usually conveyed in the form of the steady-state characteristics, which indicate how the motor behaves when any transient effects (caused for example by a sudden change in the load) have died away and conditions have once again become steady. Steady-state characteristics are usually much easier to predict than transient characteristics, and for the d.c. machine they can all be deduced from the simple equivalent circuit in Figure 3.6.

Under steady conditions, the armature current I is constant and equation (3.7) simplifies to

$$V = E + IR \qquad (3.8a)$$

or

$$I = (V - E)/R \qquad (3.8b)$$

We will derive the steady-state torque-speed characteristics for any given armature voltage V, but first we begin by establishing the relationship between the no-load speed and the armature voltage, since this is the foundation on which the speed control philosophy is based.

No-load speed

By 'no-load' we mean that the motor is running light, so that the only mechanical resistance is that due to its own friction. In any sensible motor the frictional torque will be small, and only a small driving torque will therefore be needed to keep the motor running. Since motor torque is proportional to current (equation 3.1), the no-load current will also be small. If we assume that the no-load current is in fact zero, the calculation of no-load speed becomes very simple. We note from equation 3.8b that zero current implies that the back e.m.f. is equal to the applied voltage, while equation 3.6 shows that the back e.m.f. is proportional to speed. Hence under true no-load (zero torque) conditions,

$$V = E = K_E \phi n \qquad (3.9a)$$

or

$$n = \frac{V}{K_E \phi}. \qquad (3.9b)$$

At this stage we are concentrating on the steady-state running speeds, but we are bound to wonder how it is that the motor reaches speed from rest. We will return to this when we look at transient behaviour, so for the moment it is sufficient to recall that we came across an equation identical to equation 3.9 when we looked at the primitive motor in Chapter 1. We saw that if there was no load on the shaft, the speed would rise until the back e.m.f. equalled the supply voltage. The same result clearly applies to the real d.c. motor here.

We see from equation 3.9 that the no-load speed is directly proportional to armature voltage, and inversely proportional to field flux. For the moment we will continue to consider the case where the the flux is constant, and demonstrate by means of an example that the approximations used in arriving at equation 3.9 are justified in practice. Later, we can use the same example to study the torque-speed characteristic.

Performance calculation – example

Consider a 500 V, 10 kW, 20 A motor with an armature resistance of 1Ω. When supplied at 500 V, the unloaded motor runs at 1040 rev/min., drawing a current of 0.8 A. Note that because this is a real motor, it draws a small current (and therefore produces some torque) even when unloaded. The fact that it needs to produce torque, even though it is not accelerating, is of course attributable to the inevitable friction in the bearings and brushgear.

If we want to estimate the no-load speed at a different armature voltage, say 250 V, we would use equation 3.9, giving

no-load speed at 250 V = (250/500) × 1 040 = 520 rev/min

If we insist on being more precise, we must first calculate the original value of the back e.m.f., using equation 3.8, which gives

$$E = 500 - 0.8 \times 1 = 499.2 \text{ volts}$$

The corresponding speed is 1 040 rev/min, so the e.m.f. constant must be 499.2/1 040 or 480 volts/1 000 rev/min. To calculate the no-load speed for V = 250 volts, we must first assume that the friction torque still corresponds to an armature current of 0.8 A, in which case the back e.m.f. will be given by

$$E = 250 - 0.8 \times 1 = 249.2 \text{ volts}$$

and hence the speed will be given by

no-load speed at 250 V = (249.2/480) × 1 000 =
519.2 rev/min.

The difference between the approximate and true no-load speeds is very small, and is unlikely to be significant. Hence we can safely use equation 3.9 to predict the no-load speed at any armature voltage, and obtain the set of no-load speeds shown in Figure 3.7.

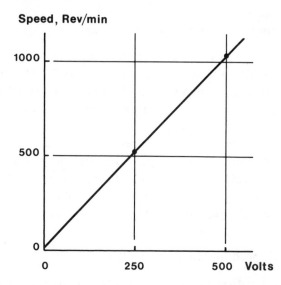

Figure 3.7   *No-load speed of d.c. motor as a function of armature voltage*

Behaviour when loaded

Having seen that the steady speed of the unloaded motor is directly proportional to the armature voltage, we need to explore how the speed will vary if we change the load on the shaft.

The usual way we quantify load is to specify the torque needed to drive the load at a particular speed. Some loads, such as a simple drum-type hoist with a constant weight on the hook, require the same torque regardless of speed, but for most loads the torque needed varies with the speed. For a fan, for example, the torque needed varies roughly with the square of the speed. If we know the torque/speed characteristic of the load, and the torque/speed characteristic of the motor, we can find the steady-state speed simply by finding the intersection of the two curves in the torque-speed plane. An example is shown in Figure 3.8.

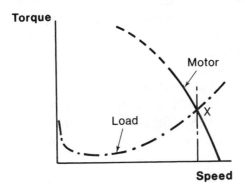

Figure 3.8   *Steady-state torque-speed curves for motor and load showing location of steady running speed*

At point X the torque produced by the motor is exactly equal to the torque needed to keep the load turning, so the motor and load are in equilibrium and the speed remains steady. At all lower speeds, the motor torque is higher than the load torque, so the nett torque will be positive, leading to an acceleration of the motor. As the speed rises towards X the acceleration reduces until the speed stabilizes at X. Conversely, at speeds above X the motor's driving torque is less than the braking torque exerted by the load, so the nett torque is negative and the system will decelerate until it reaches equilibrium at X. This example is one which is inherently stable, so that if the speed is disturbed for some reason from the point X, it will always return there when the disturbance is removed.

Turning now to the derivation of the torque/speed characteristics of the d.c. motor, we can profitably use the example above to illustrate matters. We can obtain the full-load speed for V = 500 volts by first calculating the back e.m.f. at full load (i.e. when the current is 20 A). From equation 3.8 we obtain

$$E = 500 - 20 \times = 480 \text{ volts}.$$

We have already seen that the e.m.f. constant is 480 volts/1 000 rev/min, so the full load speed is clearly 1 000 rev/min.

From no-load to full-load the speed falls linearly, giving the torque-speed curve for V = 500 volts shown in Figure 3.9. Note that from no-load to full-load the speed falls from 1 040 rev/min to 1 000 rev/min, a drop of only 4 per cent. Over the same range the back e.m.f. falls from very nearly 500 volts to 480 volts, which of course also represents a drop of 4 per cent.

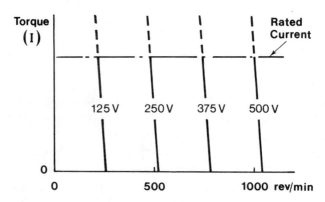

Figure 3.9   *Family of steady-state torque-speed curves for a range of armature voltages*

Two important observations follow from these calculations. Firstly, the speed drop with load is very small. This is very desirable for most applications, since all we have to do to maintain almost constant speed is to set the appropriate armature voltage and keep it constant. Secondly, a delicate balance between V and E is revealed. The current is in fact proportional to the difference between V and E, so that quite small changes in either V or E give rise to disproportionately large changes in the current. In the example, a 4 per cent reduction in E causes the current to rise to its rated value. Hence to avoid excessive currents (which cannot be tolerated in a thyristor supply, for example), the difference between V and E must be limited. This point will be taken up again when transient performance is explored.

A representative family of torque-speed characteristics for the motor discussed above is shown in Figure 3.9. As already explained, the no-load speeds are proportional to the applied voltage, while the slope of each curve is the same, being determined by the armature resistance: the smaller the resistance the less the speed falls with load. These operating characteristics are very attractive because the speed can be set simply by applying the correct voltage.

The upper region of each characteristic in Figure 3.9 is shown dotted because in this region the armature current is above its rated value, and the motor cannot therefore be operated continuously without overheating. Motors can and do operate for short periods above rated current, and the fact that the d.c. machine can continue to provide torque in proportion to current well into the overload region makes it particularly well-suited to applications requiring the occasional boost of excess torque.

A cooling problem might be expected when motors are run continuously at full current (i.e. full torque) even at very low speed, where the natural ventilation is poor. This operating condition is considered quite normal in converter-fed motor drive systems, and motors are accordingly fitted with a small blower motor as standard.

Base speed and field weakening

The speed corresponding to full armature voltage and full flux is known as base speed (see Figure 3.10).

The motor can operate at any speed up to base speed, and any torque (current) up to rated value by appropriate choice of armature voltage. This full flux region of operation is indicated by the area Oabc in Figure 3.10, and is often

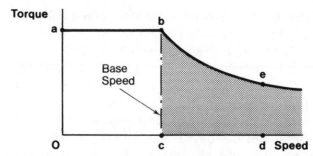

Figure 3.10   *Region of steady-state operation in the torque-speed plane*

referred to as the 'constant torque' region. In this context constant torque signifies that at any speed below base speed the motor can be loaded up to its full rated torque (current). When the current is at maximum (i.e. along the line ab in Figure 3.10), the power output is proportional to the speed, and the maximum power thus corresponds to the point b in Figure 3.10. At point b both the voltage and current have their full rated values.

To run faster than base speed the field flux must be reduced, as indicated by equation 3.9b. Operation with reduced flux is known as 'field weakening'. For example by halving the flux (and keeping the armature voltage at its full value), the no-load speed is doubled (point d in Figure 3.10). The increase in speed is however obtained at the expense of available torque, which is proportional to flux times current (equation 3.1). The current is limited to rated value, so if the flux is halved, the speed will double but the maximum torque which can be developed is only half the rated value (point e in Figure 3.10). Note that at the point e both the armature voltage and the armature current have their full rated values, so the power is at maximum, as it was at point b. The power is constant along the curve through b and e, and for this reason the shaded area to the right of the line bc is referred to as the 'constant power' region. Obviously, field weakening is only satisfactory for applications which do not demand full torque at high speeds, such as electric traction.

The maximum allowable speed under weak field conditions must be limited (to avoid excessive sparking at the commutator), and is usually indicated on the motor rating plate. A marking of 1 200/1 750 rev/min, for example, would indicate a base speed of 1 200 rev/min, and a maximum speed with field weakening of 1 750 rev/min. The field weakening range varies widely depending on the motor design, but maximum speed rarely exceeds three or four times base speed.

To sum up, the speed is controlled as follows:

- Below base speed, the flux is maximum, and the speed is set by the armature voltage. Full torque is available at any speed.
- Above base speed the armature voltage is at (or close to) maximum, and the flux is reduced in order to raise the speed. The maximum torque available reduces in proportion to the flux.

TRANSIENT BEHAVIOUR – CURRENT SURGES

It has already been pointed out that the steady-state armature current depends on the small difference between the back e.m.f. E and the applied voltage V. In a converter-fed drive it is vital that the current is kept within safe bounds, otherwise the thyristors or transistors (which have very limited overcurrent capacity) will be destroyed, and it follows from equation 3.8 that we cannot afford to let V and E differ by more than IR, where I is the rated current.

It would be unacceptable, for example, to attempt to bring a motor up to speed simply by switching on rated voltage. In the example studied earlier, rated voltage is 500 V, and the armature resistance is 1 Ω. At standstill the back e.m.f. is zero, and hence the initial current would be 500/1 = 500 A, or 25 times rated current! This would destroy the thyristors (or rather blow the fuses). Clearly the initial voltage we must apply is much less than 500 V; and if we want to limit the current to rated value (20 A in the example) the voltage needed will be 20 × 1, i.e. only 20 volts. As the speed picks up, the back e.m.f. rises, and to maintain the full current V must also be ramped up so that the difference between V and E remains constant at 20 V. Of course, the motor will not accelerate nearly so rapidly when the current is kept in check as it would if we had switched on full voltage, and allowed the current to do as it pleased. But this is the price we must pay in order to protect the converter.

Similar current-surge difficulties occur if the load on the motor is suddenly increased, because this will result in the motor slowing down, with a consequent fall in E. In a sense we welcome the fall in E because this brings about the increase in current needed to supply the extra load, but of course we only want the current to rise to its rated value; beyond that point we must be ready to reduce V, to prevent an excessive current.

Plate 3.2   *Totally-enclosed fan-ventilated wound-field d.c. motors. The smaller motor is rated at 500 W at 1500 rev/min, while the larger is rated at 10 kW at 2000 rev/min. Both motors have finned aluminium frames* (*Photograph by courtesy of GEC Electromotors Ltd*)

The solution to the problem of overcurrents lies in providing closed-loop current-limiting as an integral feature of the motor/drive package. The motor current is sensed, and the voltage V is automatically adjusted so that rated current is either never exceeded or is allowed to reach perhaps twice rated value for a few seconds. We will discuss the current control loop in Chapter 4.

SHUNT, SERIES AND COMPOUND MOTORS

Before variable-voltage supplies became readily available, most d.c. motors were obliged to operate from a single d.c. supply, usually of constant voltage. The armature and field circuits were therefore designed either for connection in parallel (shunt), or in series. As we will see shortly, the operating charactersitics of shunt and series machines differ widely, and hence each type tended to claim its particular niche: shunt motors were judged to be good for constant-speed applications, while series motors were (and still are) widely used for traction applications.

In a way it is unfortunate that these historical patterns of association have become so deep-rooted. The fact is that a converter-fed separately-excited motor, freed of any constraint between field and armature, can do everything that a shunt or series motor can, and more; and it is doubtful if shunt and series motors would ever have become widespread if variable-voltage supplies had always been around. Both shunt and series motors are handicapped in comparison with the separately-excited motor, and we will therefore be well advised to view their oft-proclaimed merits with this in mind.

The operating characteristics of shunt, series and compound (a mixture of both) motors are explored below, but first we should say something of the physical differences. At a fundamental level these amount to very little, but in the detail of the winding arrangement, they are considerable.

For a given continuous output power rating at a given speed, we find that shunt and series motors are the same size, with the same rotor diameter, the same poles, and the same quantities of copper in the armature and field windings. This is to be expected when we recall that the power output depends on the specific magnetic and electric loadings, so we anticipate that to do a given job, we will need the same amounts of active material.

The differences emerge when we look at the details of the windings, especially the field system, and they can best be illustrated by means of an example which contrasts shunt and series motors for the same output power.

Suppose that for the shunt version the supply voltage is 500 V, the rated armature (work) current is 50 A, and the field coils are required to provide an MMF of 500 Amp-turns. The field might typically consist of say 200 turns of wire with a total resistance of 200 Ω. When connected across the supply (500 V), the field current will be 2.5 A, and the MMF will be 500 AT, as required. The power dissipated as heat in the field will be 500 V × 2.5 A = 1.25 kW, and the total power input at rated load will be 500 V × 52.5 A = 26.25 kW.

To convert the machine into the equivalent series version, the field coils need to be made from much thicker conductor, since they have to carry the armature current of 50 A, rather than the 2.5 A of the shunt motor. So, working at the same current density, the cross-section of each turn of the series field winding needs to be twenty times that of the shunt field wires, but conversely only one-twentieth of the turns (i.e 10) are required for the same ampere-turns. The new field winding will therefore have a much lower resistance, of 0.5 Ω.

We can now calculate the power dissipated as heat in the series field. The current is 50 A, the resistance is 0.5 Ω, so the volt-drop across the series field is 25 V, and the power wasted as heat is 1.25 kW. This is the same as for the shunt machine, which is to be expected since both sets of field coils are intended to do the same job.

In order to allow for the 25 V dropped across the series field, and still meet the requirement for 500 V at the armature, the supply voltage must now be 525 V. The rated current is 50 A, so the total power input is 525 V × 50 A = 26.25 kW, the same as for the shunt machine.

This example illustrates that in terms of their energy-converting capabilities, shunt and series motors are fundamentally no different. Shunt machines usually have field windings with a large number of turns of fine wire, while series machines have a few turns of thick conductor. But the total amount and disposition of copper is the same, so the energy-converting abilities of both types are identical. In terms of their operating characteristics, however, the two types differ widely, as we will now see.

Shunt motor – steady-state operating characteristics

A basic shunt-connected motor has its armature and field in parallel across a single d.c. supply, as shown in Figure 3.11(a). Normally, the voltage will be constant and at the rated value for the motor, in which case the steady-state torque/speed curve will be similar to that of a permanent-magnet motor, i.e. the speed will drop slightly with load, as shown by the solid line in Figure 3.11(b). Over the normal operating region the torque-speed characteristic is similar to that of the induction motor, so shunt motors are suited to the same sorts of applications.

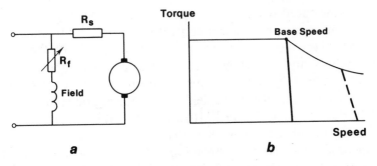

Figure 3.11   *Shunt-connected d.c. motor and steady-state torque-speed curves*

Except for small motors (say less than about 1 kW), it will be necessary to provide an external 'starting resistance' (R_s in Figure 3.11) in series with the armature, to limit the heavy current which would flow if the motor was simply switched directly onto the supply. This starting resistance is progressively reduced as the motor picks up speed, the current falling as the back e.m.f. rises. In a manual starter the resistance is controlled by the operator, while in an automatic starter the motor voltage or current are sensed and the resistance is shorted out in predetermined stages.

We should ask what happens if the supply voltage varies for any reason, and as usual the easiest thing to look at is the case where the motor is running light, in which case the back e.m.f. will almost equal the supply voltage. If we reduce the supply voltage, intuition might lead us to anticipate a fall in speed, but in fact two contrary effects occur which leave the speed almost unchanged.

If the voltage is halved, for example, both the field current and the armature voltage will be halved, and if the magnetic circuit is not saturated the flux will also halve. The new steady value of back e.m.f. will have to be half its original value, but since we now have only half as much flux, the speed will be the same. The maximum output power will of course be reduced, since at full load (i.e. full current) the power available is proportional to the armature voltage. Of course if the magnetic circuit is saturated, a modest reduction in applied voltage may cause very little drop in flux, in which case the speed will fall in proportion to the drop in voltage.

Some measure of speed control is possible by weakening the field (by means of the resistance (R_f) in series with the field winding), and this allows the speed to be raised above base value, but only at the expense of torque. A typical torque-speed charactersitic in the field-weakening region is shown by the dotted line in Figure 3.11(b). Note that it is not possible to lower the speed below the base value unless an efficient method of armature voltage control is employed, in which case a separately excited motor would almost certainly be used.

Reverse rotation is achieved by reversing the connections to either the field or the armature. The field is usually preferred since the current rating of the switch or contactor will be lower than for the armature.

Series motor – steady-state operating characteristics

The series connection of armature and field windings (Figure 3.12(a)) means that the flux is directly proportional to the current, and the torque is therefore proportional to the square of the current. Reversing the direction of the applied voltage (and hence current) therefore leaves the direction of torque unchanged. This unusual property is put to good use in the universal motor, but is a handicap when negative (braking) torque is required, since either the field or armature connections must then be reversed.

If the armature and field resistance volt-drops are neglected, and the applied voltage (V) is constant, the current varies inversely with the speed, while the torque (T) and speed (N) are related by

$$T \; a \left(\frac{V}{N}\right)^{2}. \tag{3.10}$$

A typical torque-speed characteristic is shown in Figure 3.12(b). The torque at zero speed is not infinite of course, because of the effects of saturation and resistance, both of which are ignored in equation 3.10.

It is important to note that under normal running conditions the volt-drop across the series field is only a small part of the applied voltage, most of the voltage being across the armature, in opposition to the back e.m.f. This is of course what we need to obtain efficient energy-conversion. Under starting conditions however the back e.m.f. is zero, and if the full voltage was applied the current would be excessive, being limited only by the armature and field

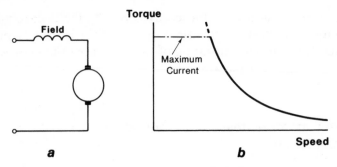

Figure 3.12   *Series-connected d.c. motor and steady-state torque-speed curve*

resistances. Hence for all but small motors a starting resistance is required to limit the current to a safe value.

Returning to Figure 3.12(b), we note that the series motor differs from most other motors in having no clearly defined no-load speed, i.e. no speed (other than infinity) at which the torque produced by the motor falls to zero. This means that, even when running light, the speed of the motor depends on the windage and friction torques, equilibrium being reached when the motor torque equals the friction torque. In large motors, the windage and friction torque is often relatively small, and the no-load speed is then too high for mechanical safety. Large series motors should therefore never be run uncoupled from their loads. As with shunt motors, the connections to either the field or armature must be reversed in order to reverse the direction of rotation.

Large series motors have traditionally been used for traction. Often books say this is because the series motor has a high starting torque which is necessary to provide acceleration of the vehicle from rest. In fact any d.c. motor of the same frame size will give the same starting torque, there being nothing special about the series motor in this respect. The real reason for its widespread use is that under the simplest possible supply arrangement (i.e. constant voltage) the overall shape of the torque-speed curve fits well with what is needed in traction applications. This was particularly important in the days when it was simply not feasible to provide for any sophisticated control of the armature voltage.

The inherent suitability of the series motor for traction is illustrated by the curves in Figure 3.13, which relate to a railway application. The solid line represents the motor characteristic, while the dotted line is the steady-state torque-speed curve for the train, i.e. the torque which the motor must provide to overcome the rolling resistance and keep the train running at each speed.

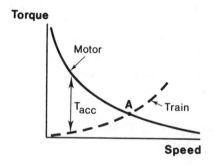

Figure 3.13   *Torque-speed curves illustrating the application of a series-connected d.c. motor to traction*

At low speeds the rolling resistance is low, the motor torque is much higher, and therefore the nett torque is large and the train accelerates at a high rate. As the speed rises, the nett torque diminishes and the acceleration tapers-off until the steady speed is reached at point A in Figure 3.13.

Some form of speed control is obviously necessary in the example above if the speed of the train is not to vary when it encounters a gradient, which will result in the rolling resistance curve shifting up or down. There are basically three methods which can be used to vary the torque-speed characteristics, and they can be combined in various ways.

Firstly, resistors can be placed in parallel with the field or armature, so that a specified fraction of the current bypasses one or the other. Field 'divert' resistors are usually preferred since their power rating is lower than armature divert resistors. For example, if a resistor with the same resistance as the field winding is switched in parallel with it, half of the armature current will now flow through the resistor and half will flow through the field. At a given speed and applied voltage, the armature current will increase substantially, so the flux will not fall as much as might be expected, and the torque will rise, as shown in Figure 3.14(a). This method is inefficient because power is wasted in the resistors, but is simple and cheap to implement. A more efficient method is to provide 'tappings' on the field winding, which allow the number of turns to be varied, but of course this can only be done if the motor has the tappings brought out.

Secondly, if a multi-cell battery is used to supply the motor, the cells may be switched progressively from parallel to series to give a range of discrete steps of motor voltage,

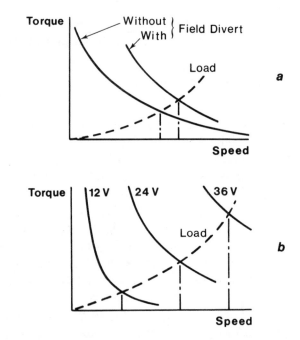

Figure 3.14 *Series motor characteristics with (a) field divert control and (b) series/parallel switching*

and hence a series of torque-speed curves. Road vehicles with 12 V lead-acid batteries often use this approach to provide say 12, 24 and 36 V for the motor, thereby giving three discrete speed settings, as shown in Figure 3.14(b).

Finally, where several motors are used (e.g. in a multiple-unit railway train) and the supply voltage is fixed, the motors themselves can be switched in various series/parallel groupings to vary the voltage applied to each.

Universal motors

In terms of numbers the main application area for the series commutator motor is in portable power tools, foodmixers, vacuum cleaners etc., where paradoxically the supply is a.c. rather than d.c. Such motors are often referred to as 'universal' motors because they can run from either a d.c. or an a.c. source.

At first sight the fact that a d.c. machine will work on a.c. is hard to believe. But when we recall that in a series motor the field flux is set up by the current which flows in the armature, it can be seen that reversal of the current will leave the direction of torque unchanged. When the motor is connnected to a 50 Hz supply for example, the (sinusoidal) current will change direction every 10 ms, and there will be a peak in the torque a hundred times per second. But the torque will always remain unidirectional, and the speed fluctuations will not be noticeable because of the smoothing effect of the armature inertia.

Series motors for use on a.c. supplies are always designed with fully laminated construction, and are intended to run at high speeds, say 8–12000 rev/min. at rated voltage. Commutation and sparking are worse than when operating from d.c., and output powers are seldom greater than 1 kW. The advantage of high speed in terms of power output per unit volume was emphasized in Chapter 1, and the universal motor is perhaps the best everyday example which demonstrates how a high power can be obtained with small size by designing for a high speed.

Until recently the universal motor offered the only relatively cheap way of reaping the benefit of high speed from single-phase a.c. supplies. Other small a.c. machines, such as induction motors and synchronous motors, were limited to maximum speeds of 3 000 rev/min. at 50 Hz (or 3 600 rev/min. at 60 Hz), and therefore could not compete in terms of power per unit volume. The availability of high-frequency inverters opens up the prospect of higher specific outputs from induction motors, and if the costs of the inverter can be brought down the universal motor will begin to face stiff competition.

Speed control of small universal motors is straight-forward using a triac (in effect a pair of thyristors connected back to back) in series with the a.c. supply. By varying the firing angle, and hence the proportion of each cycle for which the triac conducts, the voltage applied to the motor can be varied to provide speed control. This approach is widely used for electric drills, fans etc.

Compound motors

By arranging for some of the field MMF to be provided by a series winding and some to be provided by a shunt winding, it is possible to obtain motors with a wide variety of inherent torque-speed characteristics. In practice most compound motors have the bulk of the field MMF provided by a shunt field winding, so that they behave more or less like a shunt connected motor. The series winding MMF is relatively small, and is used to allow the torque-speed curve to be trimmed to meet a particular requirement.

When the series field is connected so that its MMF reinforces the shunt field MMF, the motor is said to be 'cumulatively compounded'. As. the load on the motor increases, the increased armature current in the series field causes the flux to rise, thereby increasing the torque per ampere but at the same time resulting in a bigger drop in speed as compared with a simple shunt motor. On the other hand, if the series field winding opposes the shunt winding, the motor is said to be 'differentially compounded'. In this case an increase in current results in a weakening of the flux, a reduction in the torque per ampere, but a smaller drop in speed than in a simple shunt motor. Differential compounding can therefore be used where it is important to maintain as near constant-speed as possible.

FOUR QUADRANT OPERATION AND REGENERATIVE BRAKING

One of the great beauties of the separately-excited d.c. motor is the ease with which the torque can be controlled. The torque is directly proportional to the armature current, which in turn depends on the difference between the applied voltage V and the back e.m.f. E. We can therefore make the machine run forwards or backwards, and develop positive (motoring) or negative (generating) torque simply by controlling the extent to which the applied voltage is greater or less than the back e.m.f. An armature voltage controlled d.c. machine is therefore inherently capable of what is known as '4-quadrant' operation, by reference to the torque-speed plane shown in Figure 3.15.

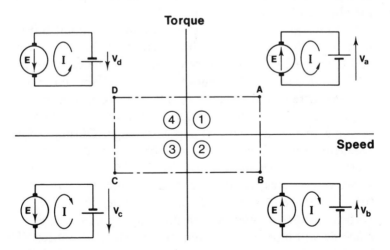

Figure 3.15 *Operation of d.c. motor in the four quadrants of the torque-speed plane*

When the machine is operating as a motor and running in the forward direction, it is operating in quadrant 1. The applied voltage V_a is greater than the back e.m.f. E, and positive current therefore flows into the motor. The power drawn from the supply (V_aI) is positive in this quadrant, and if conditions are steady most of the power will be converted and supplied to the load.

If, with the motor running at position A, we suddenly reduce the supply voltage to a value V_b which is less than the back e.m.f., the current will reverse direction, shifting the operating point to B in Figure 3.15. (If the new voltage is chosen so that $E - V_b = V_a - E$, the new current will have the same amplitude as at position A, so the new (negative) torque will be the same as the original positive torque, as shown in Figure 3.15.) But now power is supplied from the machine to the supply, i.e. the machine is acting as a generator.

We should be quite clear that all that was necessary to accomplish this remarkable reversal of power flow was a modest reduction of the voltage applied to the machine. At position A, the applied voltage was $E + IR$, while at position B it is $E - IR$. Since IR will be small compared with E, the change (2IR) is also small.

Needless to say the motor will not remain at point B if left to its own devices. The combined effect of the load torque and the negative machine torque will cause the speed to fall, so that the back e.m.f. again falls below the applied voltage V_b, the current and torque become positive again, and the motor settles back into quadrant 1, at a lower speed corresponding to the new (lower) supply voltage. During the deceleration phase, kinetic energy from the motor and load inertias is returned to the supply. This is therefore an example of regenerative braking, and it occurs naturally every time we reduce the voltage in order to lower the speed.

If we want to operate continuously at position B, the machine will have to be driven by a mechanical source. We have seen above that the natural tendency of the machine is to run at a lower speed than that corresponding to point B, so we must force it to run faster, and create an e.m.f greater than V_b if we wish it to generate continuously.

It should be obvious that similar arguments to those set out above apply when the motor is running in reverse (i.e. V is negative). Motoring then takes place in quadrant 3 (point C), with brief excursions into quadrant 4 (point D, accompanied by regenerative braking) whenever the voltage is reduced in order to lower the speed.

Full speed regenerative reversal

To illustrate more fully just how the voltage has to be varied during sustained regenerative braking, we can consider how to change the speed of an unloaded permanent-magnet motor from full speed in one direction to full speed in the other, in the shortest possible time.

At full forward speed the applied voltage is taken to be $+V$, and since the motor is unloaded the no-load current will be very small and the back e.m.f. will be almost equal to V. Ultimately, we will clearly need an armature voltage of $-V$ to make the motor run at full speed in reverse. But we cannot simply reverse the applied voltage: if we did, the armature current immediately afterwards would be given by $(-V -E)/R$, which would be disastrously high. (The motor might tolerate it for the short period for which it would last, but the supply certainly could not!)

What we need to do is adjust the voltage so that the current is always limited to rated value, and in the right direction. Since we want to decelerate as fast as possible, we must aim to keep the current negative, and at rated value throughout the period of deceleration and for the run up to full speed in reverse. This will give us constant torque throughout, so the deceleration (and subsequent acceleration) will be constant and the speed will change at a uniform rate, as shown in Figure 3.16(a).

We note that to begin with the applied voltage has to be reduced to less than the back e.m.f. and then ramped down linearly with time so that the difference between V and E is kept constant, thereby keeping the current constant at its

rated value. During the reverse run-up, V has to be
numerically greater than E, as shown in Figure 3.16(a). (The
difference between V and E has been exaggerated in Figure
3.16 for clarity: in a large motor, the difference may only be
one or two per cent at full speed.)

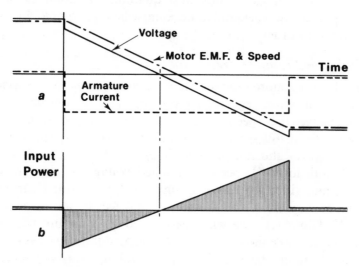

Figure 3.16 *Regenerative reversal of d.c. motor from full-speed forward to
full-speed reverse, at maximum allowable torque (current)*

The power to and from the supply is shown in Figure
3.16b, the energy being represented by the shaded areas. During the deceleration period most of the kinetic energy of the
motor (lower shaded area) is progressively returned to the
supply, the motor acting as a generator for the whole of this
time. The total energy recovered in this way can be appreciable in the case of a large drive such as a steel rolling mill.
A similar quantity of energy (upper shaded area) is supplied
and stored as kinetic energy as the motor picks up speed in
the reverse sense.

Three final points need to be emphasized. First, we have
assumed throughout the discussion that the supply can provide positive or negative voltages, and can accept positive or
negative currents. A note of caution is therefore appropriate,

because many simple power electronic converters do not have this flexibility. Users need to be aware that if full 4-quadrant operation (or even 2-quadrant regeneration) is called for, a basic converter will probably not be adequate. This point is taken up again in Chapter 4. Secondly, we should not run away with the idea that in order to carry out the reversal in Figure 3.16 we would have to work out in advance how to profile the applied voltage as a function of time. Our drive system will normally have the facility for automatically operating the motor in constant-current mode, and all we will have to do is to tell it the new target speed. This is also taken up in Chapter 4. And finally, we must remember that the discussion above relates to separately-excited motors. If regenerative braking is required for a series motor, the connections to either the field or armature must be reversed in order to reverse the direction of torque.

Dynamic braking

A simpler and cheaper, if less effective, method of braking can be achieved by dissipating the kinetic energy of the motor and load in a resistor, rather than returning it to the supply. A version of this technique is employed in the cheaper power electronic converter drives, which have no facility for returning power to the mains.

When the motor is to be stopped, the supply is removed and a resistor is switched across the armature brushes. The motor e.m.f. drives a (negative) current throuth the resistor, and the negative torque results in deceleration. As the speed falls, so does the e.m.f., the current, and the braking torque. At very low speeds the braking torque is therefore very small. Ultimately, all the kinetic energy is converted to heat in the motor's own armature resistance and the external resistance. Very rapid initial braking is obtained by using a low resistance (or even simply short-circuiting the armature). Dynamic braking is still widely used in traction because of its simplicity, though most new rapid transit schemes employ the more energy-efficient regenerative braking process.

4

D.C. MOTOR DRIVES

INTRODUCTION

In terms of numbers of installations, the controlled rectifier ('Thyristor') drive remains by far the most important form of general purpose industrial drive, so we consider it first, and in some depth. Later, we will look briefly at chopper-fed drives which are used mainly in the medium and small sizes, and finally turn attention to small servo-type drives.

THYRISTOR D.C. DRIVES

The overall arrangement of a drive intended to provide closed-loop speed control is shown in Figure 4.1.

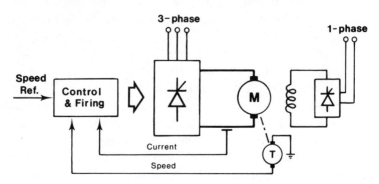

Figure 4.1   *Schematic diagram of speed-controlled d.c. motor drive*

The main power circuit consists of a six-thyristor bridge circuit (as discussed in Chapter 2) which rectifies the incoming a.c. supply to produce a d.c. supply to the motor armature. The assembly of thyristors, mounted on a heatsink, is usually referred to as the 'stack'. By altering the firing angle of the thyristors the mean value of the rectified voltage can be varied, thereby allowing the motor speed to be controlled.

For motors up to a few kilowatts the armature stack can be supplied from either single-phase or three-phase mains, but for larger motors three-phase is always used. A separate thyristor or diode rectifier is used to supply the field of the motor: the power is much less than the armature power, so the field stack is comparatively small, and the supply is often single-phase.

We saw in Chapter 2 that the controlled rectifier produces

Plate 4.1   *Force-ventilated d.c. wound-field d.c. motor, 100 kW at 2000 rev/min. The motor is of all-laminated construction, and is designed for use with a thyristor converter. The blower motor is a small single-phase induction machine, and is intended to run continuously (Photograph by courtesy of GEC Electromotors Ltd)*

a crude form of d.c. with a pronounced ripple in the output voltage. This ripple component gives rise to pulsating currents and fluxes in the motor, and in order to avoid excessive eddy-current losses and commutation problems the poles and frame should be of laminated construction. It is accepted practice for motors supplied for use with thyristor drives to have laminated construction, but older motors often have solid poles and/or frames, and these will not always work satisfactorily with a rectifier supply. It is also the norm for drive motors to be supplied with an attached blower motor as standard. This provides continuous through-ventilation and allows the motor to operate continuously at full torque even down to the lowest speeds without overheating.

Low power control circuits are used to monitor the principal variables of interest (usually motor current and speed), and to generate appropriate firing pulses so that the motor maintains constant speed despite variations in the load. The 'Speed Reference' (Figure 4.1) is usually an analogue voltage, typically varying from 0–10 V, and obtained from a manual speed-setting potentiometer or from elsewhere in the plant.

The combination of power, control and protective circuits constitutes the converter. Standard modular converters are available as off-the-shelf items in sizes from 0.5 kW up to several hundred kW, while larger drives will be tailored to individual requirements. Individual converters may be mounted in enclosures with isolators, fuses etc. or groups of converters may be mounted together to form a multi-motor drive.

Motor operation with converter supply

The basic operation of the rectifying bridge has been discussed in Chapter 2, and we now turn to the matter of how the d.c. motor behaves when supplied with d.c. from a controlled rectifier.

By no stretch of the imagination could the waveforms of

armature voltage looked at in Chapter 2 (e.g. Figure 2.11) be thought of as good d.c., and it would not be unreasonable to question the wisdom of feeding such an unpleasant-looking waveform to a d.c. motor. In fact it turns out that the motor works almost as well as it would if fed with pure d.c., for two main reasons. Firstly, the armature inductance of the motor causes the waveform of armature current to be much smoother than the waveform of armature voltage, which in turn means that the torque ripple is much less than might have been feared. And secondly, the inertia of the armature is sufficiently large for the speed to remain almost steady despite the torque ripple. It is indeed fortunate that such a simple arrangement works so well, because any attempt to smooth out the voltage waveform (perhaps by adding smoothing capacitors) would prove to be prohibitively expensive in the power ranges of interest.

Motor current waveforms

For the sake of simplicity we will look at operation from a single-phase (2-pulse) converter, but similar conclusions apply to the 6-pulse one. The voltage (V_a) applied to the motor armature is typically as shown in Figure 4.2(a).

It can be considered to consist of a mean d.c. level (V_{dc}), and a superimposed pulsating or ripple component which we can denote loosely as v_{ac}. V_{dc} can be altered by varying the firing angle, which also incidentally alters the ripple.

The ripple voltage causes a ripple current to flow in the armature, but because of the armature inductance, the amplitude of the ripple current is small. In other words, the armature presents a high impedance to a.c. voltages. This smoothing effect of the armature inductance is shown in Figure 4.2(b), from which it can be seen that the current ripple is relatively small in comparison with the corresponding voltage ripple. The average value of the ripple current is of course zero, so it has no effect on the average torque of the motor. There is a nevertheless a variation in torque every half-cycle of the mains, but because it is of small amplitude

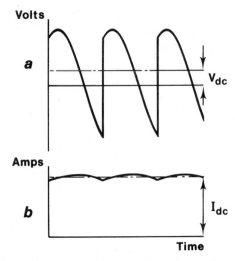

Figure 4.2   *Armature voltage (a) and current (b) for a d.c. motor supplied from a single-phase fully-controlled thyristor converter*

and high frequency the variation in speed (and hence back e.m.f., E) will not usually be noticeable.

The mean d.c. voltage and the mean d.c. current (I_{dc}) are related by the equation

$$V_{dc} = E + I_{dc}R \qquad (4.1)$$

which is exactly the same as for operation from a pure d.c. supply. In other words, the torque (I_{dc}) is governed by the mean d.c. voltage, and is independent of the ripple. This is very important, as it underlines the fact that we can control the mean motor voltage, and hence the speed, simply by varying the converter delay angle.

The smoothing effect of the armature inductance is in fact the key to successful motor operation. The armature acts as a low-pass filter, blocking most of the ripple, and leading to a more or less constant armature current. For the smoothing to be effective, the armature time-constant needs to be long compared with the pulse duration (half a cycle with a 2-pulse drive, but only one sixth of a cycle in a 6-pulse drive). This condition is met in all 6-pulse drives, and in most 2-pulse ones. Overall, the motor then behaves much as it would with pure d.c., though the I^2R loss is higher than it would be if the current was perfectly smooth.

In some small 2-pulse drives, however, there is insufficient armature inductance to prevent the current from falling to zero for part of each cycle. This is known as the 'discontinuous current' mode, and it is usually encountered when the motor is lightly loaded. It is very undesirable because when the current is discontinuous, the mean voltage from the converter falls dramatically as the mean current drawn by the motor is increased, and in addition the I^2R loss is much higher than it would be with pure d.c. current. With discontinuous current, the relevant part of the torque-speed curve is very droopy, as shown in Figure 4.3; and it may be worthwhile to add extra inductance in series with the armature in order to promote continuous current, and improve the torque-speed curve as indicated in Figure 4.3.

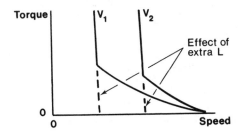

Figure 4.3   *Torque-speed curves illustrating the undesirable 'droopy' characteristic associated with discontinuous current. The improved characteristic (shown dotted) corresponds to operation with continuous current*

To conclude this section, we should also reiterate that a further 'smoothing' effect always takes place in regard to the motor speed. Even though the current (and hence the torque) varies somewhat over a pulse, the corresponding fluctuation in speed is much less because the electromechanical time-constant is invariably much longer than the pulse duration. The assumption of constant speed (and hence constant E) made earlier is thus seen to be justified.

Converter output impedance: overlap

In Chapter 2 we tacitly assumed that the output voltage from the converter was independent of the current drawn by the motor, and depended only on the delay angle a. In other words we have treated the converter as an ideal voltage source.

In practice the a.c. supply has a finite impedance, and we must therefore expect a volt-drop which depends on the motor current being drawn. Perhaps surprisingly, the supply impedance (which is mainly due to leakage reactances in transformers) manifests itself at the output stage of the converter as a supply resistance, so the supply volt-drop (or regulation) is directly proportional to the motor armature current.

It is not appropriate to go into more detail here, but we should note that the effect of supply reactance is to delay the transfer (or commutation) of the current between thyristors, a phenomena known as overlap. The consequence of overlap is that instead of the output voltage making an abrupt jump at the start of each pulse, there is a short period where two thyristors are conducting simultaneously. During this interval the output voltage is the mean of the voltages of the incoming and outgoing voltages, as shown typically in Figure 4.4. It is important for users to be aware that overlap is to be expected, as otherwise they may be alarmed the first time they connect an oscilloscope to the motor terminals. When the drive is connected to a 'stiff' industrial supply the overlap will only last for perhaps a few microseconds, so the 'notch' shown in Figure 4.4 would be barely visible on an oscilloscope. Books always exaggerate the width of the overlap for the sake of clarity, as in Figure 4.4. If in practice the overlap is seen to last for more than say 1 millisecond, the implication is that the supply system impedance is too high for the size of converter in question, or conversely, the converter is too big for the supply.

Returning to the practical consequences of supply impedance, we simply have to allow for the presence of an

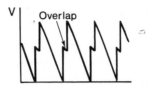

Figure 4.4   *Distortion of armature voltage waveform caused by rectifier overlap*

extra 'source resistance' in series with the output voltage of the converter. This source resistance is in series with the motor armature resistance, and hence the motor torque-speed curves for each value of a have a somewhat steeper droop than they would if the supply impedance was zero.

Four-quadrant operation and inversion

So far we have looked at the converter as a rectifier, supplying power from the a.c. mains to a d.c. machine running in the positive direction and acting as a motor. As explained in Chapter 3, this is known as 1-quadrant operation, by reference to quadrant 1 of the complete torque-speed plane shown in Figure 3.15.

But suppose we want to run as a motor in the opposite direction, with negative speed and torque, i.e. in quadrant 3. How do we do it? And what about operating the machine as a generator, so that power is returned to the a.c. supply, the converter then 'inverting' power rather than rectifying, and the system operating in quadrant 2 or quadrant 4. We need to do this if we want to achieve regenerative braking. Is it possible, and if so how?

The good news is that as we saw in Chapter 3 the d.c. machine is inherently reversible. If we apply a positive voltage V greater than E, a current flows into the armature and the machine runs as a motor. If we reduce V so that it is less than E, the current, torque and power automatically reverse direction, and the machine acts as a generator, converting mechanical energy (its own kinetic energy in the case of regenerative braking) into electrical energy. And if we want to motor or generate with the reverse direction of rotation, all we have to do is to reverse the polarity of the armature supply. The d.c. machine is inherently a 4-quadrant device, but needs a supply which can provide positive or negative voltage, and simultaneously handle either positive or negative current.

This is where we meet a snag: a single thyristor converter can only handle current in one direction, because the thyristors are unidirectional devices. This does not mean that the converter is incapable of returning power to the supply however. The d.c. current can only be positive, but (provided it is a fully-controlled converter) the d.c. output voltage can be either positive or negative. The power flow can therefore be positive (rectification) or negative (inversion).

For normal motoring where the output voltage is positive, the delay angle (α) lies in the range $0°$ to $90°$. But if α is made greater than $90°$, the output voltage goes negative, as indicated by equation 2.6, and shown in Figure 4.5. A single fully-controlled converter therefore has the potential for 2-quadrant operation, though it has to be admitted that this capability is not easily exploited unless we are prepared to employ reversing switches in the armature or field circuits. This is discussed next.

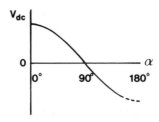

Figure 4.5 *Average d.c. output voltage from a fully-controlled thyristor converter as a function of the firing angle delay* α

Single-converter reversing drives

We will consider a fully-controlled converter supplying a permanent-magnet motor, and see how the motor can be regeneratively braked from full speed in one direction, and then accelerated up to full speed in reverse. We looked at this procedure in principle at the end of Chapter 3, but here we explore the practicalities of achieving it with a converter-fed drive. We should be clear from the outset that all the user has to do is to change the speed reference signal from full forward to full reverse: the control system in the drive converter takes care of matters from then on. What it does, and how, is discussed below.

When the motor is running at full speed forward, the converter delay angle will be small, and the converter output voltage V and current I will both be positive. This condition is shown in Figure 4.6(a), and corresponds to operation in quadrant 1.

In order to brake the motor, the torque has to be reversed. The only way this can be done is by reversing the direction of armature current. The converter can only supply positive current, however, so to reverse the motor torque we have to reverse the armature connections, using a mechanical switch or contactor, as shown in Figure 4.6(b). (Before operating the contactor, the armature current would be reduced to zero by lowering the converter voltage, so that the contactor is not required to interrupt current.) Note that because the motor is still rotating in the positive direction, the back e.m.f. remains in its original sense; but now the back e.m.f. is seen to be assisting the current and so to keep the current within bounds the converter must produce a negative voltage V which is just a little less than E. This is achieved by setting the delay angle at the appropriate point between 90° and 180°. Note that the converter current is still positive, but the converter voltage is negative, and power is thus flowing back to the mains. In this condition the system is operating in quadrant 2. As the speed falls, E reduces, and so V must be reduced progressively to keep the current at full value. This is achieved automatically by the action of the current loop, which is discussed later.

The current (i.e. torque) needs to be kept negative in order to run up to speed in the reverse direction, but after the back

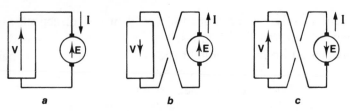

Figure 4.6   *Stages in motor reversal using a single-converter drive*

e.m.f. changes sign (as the motor reverses) the converter voltage again becomes positive and greater than E, as shown in Figure 4.6(c). The converter is then rectifying, with power being fed into the motor, and the system is operating in quadrant 3.

Schemes using reversing contactors are not suitable where the reversing time is critical, because of the delay caused by the switchover time, which may easily amount to 200–400 ms. Field reversal schemes operate in a similar way, but reverse the field current instead of the armature current. They are even slower, because of the relatively long time-constant of the field winding.

Double converter reversing drives

Where full four-quadrant operation and rapid reversal is called for, two converters connected in anti-parallel are used, as shown in Figure 4.7. One converter supplies positive current to the motor, while the other supplies negative current.

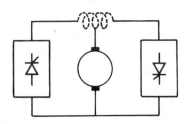

Figure 4.7   *Double-converter reversing drive*

The bridges are operated so that their d.c. voltages are almost equal thereby ensuring that any d.c. circulating current is small, and a reactor is placed between the bridges to limit the flow of ripple currents which result from the unequal ripple voltages of the two converters. Alternatively, the reactor can be dispensed with by only operating one converter at a time. The changeover from one converter to the other can only take place after the firing pulses have been removed from one converter, and the armature current has decayed to zero. Appropriate zero-current detection circuitry is of course provided as an integral part of the drive, so that as far as the user is concerned, the two converters behave as if they were a single ideal bi-directional d.c. source.

Prospective users need to be aware of the fact that a basic single converter can only provide for operation in one quadrant. If regenerative braking is required, either field or armature reversing contactors will be needed; and if rapid reversal is essential, a double converter has to be used. All these extras naturally push up the purchase price.

Power factor and supply effects

One of the drawbacks of a converter-fed d.c. drive is that the supply power-factor is very low when the motor is operating at high torque (i.e. high current) and low speed (i.e. low armature voltage), and is less than unity even at base speed and full load. This is because the supply current waveform lags the supply voltage waveform by the delay angle a, as shown in Figure 4.8, and also the supply current is approximately rectangular (rather than sinusoidal).

It is important to emphasize that the supply power-factor is always lagging, even when the converter is inverting. There is no way of avoiding the low power-factor, so users need to be prepared to augment their existing power-factor correcting equipment if necessary.

The harmonics in the mains current waveform can give rise to a variety of interference problems, and supply authorities

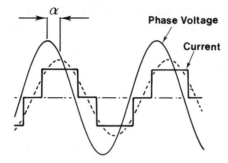

Figure 4.8 *a.c. supply voltage and current waveforms*

generally impose statutory limits. For large drives (say hundreds of kW), filters may have to be provided to prevent these limits from being exceeded.

Since the supply impedance is never zero, there is inevitably some distortion of the mains voltage waveform. This takes the form of notches in the voltage, as indicated in Figure 4.9. The notches arise because the mains is momentarily short-circuited each time the current commutates from one thyristor to the next, i.e. during the overlap period discussed earlier. For the majority of small and medium drives, connected to 'stiff' industrial supplies, these notches are too small to be noticed (they are greatly exaggerated for the sake of clarity in Figure 4.9); but they can pose a serious problem when a large drive is connected to a weak supply.

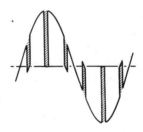

Figure 4.9 *Distortion of supply voltage waveform caused by rectifier overlap (the width of the notches has been exaggerated for the sake of clarity)*

CONTROL ARRANGEMENTS FOR D.C. DRIVES

The most common arrangement, which is used with only minor variations over the whole size range from small servo-type drives up to the largest industrial drives, is the so-called two-loop control. This has an inner feedback loop to control the current (and hence torque) and an outer loop to control speed. When position control is called for, a further outer position loop is added.

A two-loop scheme for a thyristor d.c. drive is discussed first, but the essential features are the same in a chopper-fed drive. Later the simpler arrangements used in low-cost small drives are discussed.

As far as possible the discussion is limited to those aspects which the user needs to know about and understand. In pratice, once a drive has been commissioned, there are only a few potentiometer adjustments to which the user has access. Whilst most of them are self-explanatory (e.g. max. speed, min. speed, accel. and decel. rates), some are less obvious (e.g. 'current stability', 'speed stability', 'IR' comp.) so these are explained.

Current control

A standard d.c. drive system with speed and current control is shown in Figure 4.10. At the heart of the system is the closed-loop current controller, or current loop, enclosed by the dotted lines. The purpose of the current loop is to make the actual motor current follow the current reference signal (I_{ref}) shown in Figure 4.10. It does this by comparing a feedback signal of actual motor current with the current reference signal, amplifying the difference (or current error), and using the resulting amplified current error signal (an analogue voltage) to control the firing angle a – and hence the output voltage – of the converter. The current feedback signal is obtained either from a d.c. current transformer (which gives an isolated analogue voltage output), or from a.c. current transformer/rectifiers in the mains supply lines.

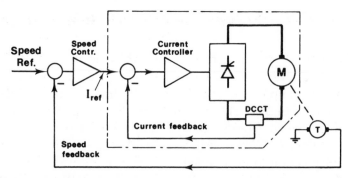

Figure 4.10   *Schematic diagram of analogue controlled-speed drive with current and speed feedback*

The job of comparing the reference (demand) and actual current signals and amplifying the error signal is carried out by the current-error amplifier. By giving the current-error amplifier a high gain, the actual motor current will always correspond closely to the current reference signal, i.e. the current-error will be small, regardless of motor speed. In other words, we can expect the actual motor current to follow the 'current reference' signal at all times, the armature voltage being automatically adjusted by the controller so that, regardless of the speed of the motor, the current has the correct value.

Of course no control system can be perfect, but it is usual for the current-error amplifier to be of the proportional plus integral (PI) type (see below), in which case the actual and demanded currents will be exactly equal under steady-state conditions.

The importance of preventing excessive currents from flowing has been emphasized previously, and the current control loop provides the means to this end. As long as the current control loop functions properly, the motor current can never exceed the reference value. Hence by limiting the magnitude of the current reference signal (by means of a clamping circuit), the motor current can never exceed the specified value. This 'electronic current limiting' is by far the most important protective feature of any drive. It means that if for example the motor suddenly stalls because the load seizes, the armature voltage will automatically reduce to a very low value, thereby limiting the current to its maximum allowable level.

The first thing we should aim at when setting up a drive is a good current loop. In this context, 'good' means that the steady-state motor current should correspond exactly with the current reference, and the transient response to step changes in the current reference should be fast and well damped. The first of these requirements is satisfied by the integral term in the current-error amplifier, while the second is obtained by judicious choice of the amplifier gain and time-constant. As far as the user is concerned, the 'current stability pot' is provided to allow him to optimize the transient response of the current loop.

On a point of jargon, it should perhaps be mentioned that the current-error amplifier is more often than not called either the 'current controller' or the 'current amplifier'. The first of these terms is quite sensible, but the second can be very misleading: there is after all no question of the motor current itself being amplified.

Torque control

For applications requiring the motor to operate with a specified torque regardless of speed (e.g. in line tensioning), we can dispense with the outer (speed) loop, and simply feed a current reference signal directly to the current controller (usually via the 'torque ref' terminal on the control board). This is because torque is directly proportional to current, so the current controller is in effect also a torque controller. We may have to make an allowance for accelerating torque, by means of a transient 'inertia compensating' signal, but this is usually provided for via a potentiometer adjustment.

In the current-control mode the current remains constant at the set value, and the steady running speed is determined by the load. If the torque reference signal was set at 50 per cent, for example, and the motor was initially at rest, it would accelerate with a constant current of half rated value until the load torque was equal to the motor torque. Of course, if the motor was running without any load, it would accelerate quickly, the applied voltage ramping up so that it always remained higher than the back e.m.f. by the amount needed to drive the specified current into the armature. Eventually the motor would reach a speed (a little above normal full speed) at which the converter output voltage had reached its upper limit, and it was therefore no longer possible to maintain the set current.

Speed control

The outer loop in Figure 4.10 provides speed control. Speed feedback is provided by a d.c. tachogenerator, and the actual and required speeds are fed into the speed-error amplifier (often known simply as the speed amplifier).

Any difference between the actual and desired speed is amplified, and the output serves as the input to the current loop. Hence if for example the actual motor speed is less than the desired speed, the speed amplifier will demand current in proportion to the speed error, and the motor will therefore accelerate in an attempt to minimize the speed error. As the speed comes up to target, the speed error reduces, and the final speed is approached smoothly.

In this mode the speed will be held at the value set by the speed reference signal for all loads up to the point where full armature current is needed. If the load torque increases any more the speed will drop because the current-loop will not allow any more armature current to flow. Conversely, if the load attempted to force the speed above the set value, the motor current will be reversed automatically, so that the motor acts as a brake and regenerates power to the mains.

The speed-error amplifier will usually be of PI (proportional plus integral) type, so that the steady-state speed error will be zero. As mentioned above, the current reference signal is clamped at a level corresponding to a safe maximum current. This means that beyond a certain speed error, no extra current will be permitted, no matter how far off target the speed is. When starting from rest, for example, we can safely apply a step demand for full speed. The output (I_{ref}) from the speed-error amplifier will immediately saturate at its maximum value, which is deliberately clamped so as to correspond to a demand for the maximum (rated) current in the motor. The motor current will therefore be at rated value, and the motor will accelerate at full torque. (In some drives the current reference is allowed to reach 150 per cent or even 200 per cent of rated value for a few seconds, in order to provide a short torque-boost. This is particularly valuable in starting loads with high static friction, and is known as 'two-stage current limit'.)

The output of the speed amplifier will remain saturated until the actual speed is quite close to the target speed, and for all this time the motor current will therefore be held at full value. Only when the speed is within a few per cent of target will the speed amplifier come out of saturation. Thereafter, as the speed continues to rise, and the speed error falls, the output of the speed amplifier falls below the clamped level. Speed control then enters a linear regime, in which the correcting current (and hence the torque) is proportional to speed error, giving a smooth approach to final speed.

A good speed controller will result in zero steady-state error, and have a well-damped response to step changes in the demanded speed. The integral term in the PI control caters for the requirement of zero steady-state error, while the transient response depends on the setting of the controller gain and time-constant. The speed stability potentiometer is provided to allow the user to optimize the transient speed response.

It should be noted that it is generally much easier to obtain a good transient response with a regenerative drive, which has the ability to supply negative current (i.e. braking torque) should the motor overshoot the desired speed. A non-regenerative drive cannot furnish negative current (unless fitted with reversing contactors), so if the speed overshoots the target the best that can be done is to wait for the motor to decelerate naturally. This is not satisfactory, and every effort therefore has to be made to avoid controller settings which lead to an overshoot of the target speed.

As with any closed-loop scheme, problems occur if the feedback signal is lost when the system is in operation. If the tacho feedback became disconnected, the speed amplifier would immediately saturate, causing full torque to be applied. The speed would then rise until the converter output reached its maximum output voltage. To guard against this many drives incorporate tacho-loss detection circuitry, and in some cases armature voltage feedback (see later section) automatically takes over in the event of tacho failure.

Drives which use field-weakening include automatic provision for controlling both armature voltage and field current when running above base speed. Typically, the field current is kept at full value until the armature voltage reaches about 95 per cent of rated value. When a higher speed is demanded, the extra armature voltage applied is accompanied by a simultaneous reduction in the field current, in such a way that when the armature voltage reaches 100 per cent the field current is at the minimum safe value. This process is known as 'spillover field weakening'.

Overall operating region

A standard drive with field-weakening provides armature voltage control of speed up to base speed, and field-weakening control of speed thereafter. Any torque up to the rated value can be obtained at any speed below base speed, and as explained in Chapter 3 this region is known as the 'constant torque' region. Above base speed, the maximum available torque reduces inversely with speed, so this is known as the 'constant power' region. For a converter-fed drive the operating region in quadrant 1 of the torque-speed plane is therefore as shown in Figure 3.10. (If the drive is equipped for regenerative and reversing operation, the operating area is mirrored in the other three quadrants, of course.)

Armature voltage feedback and IR compensation

In low-power drives where precision speed-holding is not essential, and cost must be kept to a minimum, the tachogenerator is dispensed with and the armature voltage is used as a 'speed feedback' instead. Performance is clearly not as good as with tacho feedback, since whilst the steady-state no-load speed is proportional to armature voltage, the speed falls as the load (and hence armature currrent) increases.

We saw in Chapter 3 that the drop in speed with load was attributable to the armature resistance volt-drop (IR), and the drop in speed can therefore be compensated by boosting the applied voltage in proportion to the current. A potentiometer adjustment labelled 'IR comp' or simply 'IR' is provided on the drive circuit for the user to adjust to suit the particular motor. The compensation is usually far from perfect, since it cannot cope with temperature variation of resistance, nor with the effects of armature reaction, but it is better than nothing.

Drives without current control

Cheaper drives often dispense with the full current control loop, and incorporate a crude but effective 'current-limit' which only operates when the maximum set current would otherwise be exceeded. These drives usually have an in-built ramp circuit which limits the rate of rise of the set speed signal so that under normal conditions the current limit is not activated.

CHOPPER-FED D.C. MOTOR DRIVES

If the source of supply is d.c. (for example in a battery vehicle or a rapid transit system) a chopper-type regulator is usually employed. The basic operation of a single-switch chopper was discussed in Chapter 2, where it was shown that the average output voltage could be varied by periodically switching the battery voltage on and off for varying intervals.

A single-switch chopper using a transistor, GTO or comparable switching device can only supply positive voltage and current to a d.c. motor, and is therefore restricted to one-quadrant motoring operation. When regenerative and/or rapid speed reversal is called for, more complex circuitry is required, involving two or more power switches, and consequently leading to increased cost. Many different circuits are used and it is not appropriate to go into detail here in view of the comparatively restricted range of applications. It should however be mentioned that while the chopper circuit discussed in Chapter 2 will only provide an ouput voltage in the range $0 < E$, where E is the battery voltage, it is possible to construct a 'step-up' chopper which will deliver d.c. with an average voltage greater than the battery voltage.

Performance of chopper-fed d.c. motor drives

We saw earlier that the d.c motor performed almost as well when fed from a phase-controlled rectifier as it does when supplied with pure d.c. The chopper-fed motor is, if anything, rather better than the phase-controlled, because the armature voltage ripple is generally less. Typical waveforms of armature voltage and current are shown in Figure 4.11.

These can be compared with the phase-controlled equivalents shown in Figure 4.2, and it will be clear that for the same mean current, the chopper-fed motor has a rather lower ripple. Naturally, a loose comparison like this is only valid for a given chopping frequency, and in general the higher the frequency, the lower the ripple. A frequency of around 100 Hz, as shown in Figure 4.11, is however typical of a medium-sized chopper drive.

Torque-speed and control arrangements

Under open-loop conditions (i.e. where the mark-space ratio of the chopper is fixed at a particular value) the behaviour of the chopper-fed motor is similar to the converter-fed motor

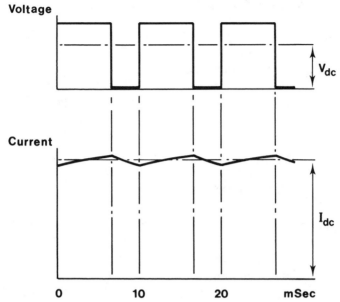

Figure 4.11 *Typical armature voltage and current waveforms for a chopper-fed d.c. motor*

discussed earlier (see Figure 4.3). When the armature current is continuous the speed falls only slightly with load, because the armature voltage remains constant. But when the armature current is discontinuous (which is most likely at high speeds and light load) the speed falls off rapidly when the load increases, because the armature voltage changes with load. Discontinuous current can be avoided by adding an inductor in series with the armature, or by raising the chopping frequency, but when closed-loop speed control is employed, the undesirable effects of continuous current are masked by the control loop.

The control philosophy and arrangements for a chopper-fed motor are the same as for the converter-fed motor, with the obvious exception that the mark-space ratio of the chopper is used to vary the output voltage, rather than the firing angle.

DEVELOPMENTS IN INDUSTRIAL D.C. DRIVES

Some drives now feature digital speed feedback, in which a pulse train generated from a shaft-mounted encoder is compared (using a phase-locked-loop) with a reference pulse train whose frequency corresponds to the desired speed. The reference frequency can easily be made accurate and drift-free; and noise in the encoder signal is easily rejected, so that very precise speed holding can be guaranteed. This is especially important when a number of independent motors must all be driven at identical speed. Phase-locked loops are also used in the firing-pulse synchronizing circuits, to overcome the problems caused by noise on the mains waveform.

Digital controllers are steadily taking over from the analogue amplifiers discussed earlier. The advantages are freedom from drift, added flexibility in relation to the control stategies which can be employed, ease of interfacing to host computers and controllers, and, in the longer term, the prospect of self-tuning.

User-friendly diagnostics represents another rapidly growing area of activity, which promises to provide the user with current and historical data on the state of all the key drive variables.

D.C. SERVO DRIVES

The precise meaning of the term servo in the context of motors and drives is difficult to pin down. Broadly speaking, if a drive incorporates servo in its description, the implication is that it is intended specifically for closed-loop or feedback control, usually of shaft torque, speed or position. Early servomechanisms were developed primarily for military applications, and it quickly became apparent that standard d.c. motors were not always suited to precision control. In particular high torque to inertia ratios were needed, together with smooth ripple-free torque. Motors were therefore developed to meet these exacting requirements, and not surprisingly they were, and still are, much more expensive than

Plate 4.2   *Microprocessor-controlled d.c. motor drive, available for nominal power ratings up to 575 kW for both single-quadrant or four-quadrant drives. The programmer allows online adjustment of drive parameters (maximum and minimum speed, torque limit etc.), as well as providing monitoring and pre-fault record facilities (Photograph by courtesy of GEC Industrial Controls Ltd)*

their industrial counterparts. Whether the extra expense of a servo motor can be justified depends on the specification, but prospective users should always be on their guard to ensure they are not pressed into an expensive purchase when a conventional industrial drive could cope perfectly well.

The majority of servo drives are sold in modular form, consisting of a high-performance permanent magnet motor, often with an integral tachogenerator, and a chopper-type power amplifier module. The drive amplifier normally requires a separate regulated d.c. power supply, if, as is normally the case, the power is to be drawn from the a.c. mains. Continuous output powers range from a few watts up to perhaps 2–3 kW, with voltages of 12 V, 24 V, 48 V and multiples of 50 V being standard.

Servo motors

Although there is no sharp dividing line between servo motors and ordinary motors, the servo type will be intended for use in applications which require rapid acceleration and deceleration. The design of the motor will reflect this by catering for intermittent currents (and hence torques) of many times the continuously rated value. Because most servo motors are small, their armature resistances are relatively high: the short-circuit (locked-rotor) current at full armature voltage is therefore perhaps only five times the continuously rated current, and the drive amplifier will normally be selected so that it can cope with this condition without difficulty, giving the motor a very rapid acceleration from rest. The even more arduous condition in which the armature voltage is suddenly reversed with the motor running at full speed is also quite normal. (Both of these modes of operation would of course be quite unthinkable with a large d.c. motor, because of the huge currents which would flow as a result of the much lower armature resistance.)

In order to maximize acceleration, the rotor inertia must be minimized, and one obvious way to achieve this is to construct a motor in which only the electric circuit (con-

Plate 4.3 *High-performance permanent-magnet brushed d.c. servo motors. The smaller motors (sizes 15, 18 and 23) employ Samarium Cobalt magnets giving exceptionally high torque/volume ratios. In the larger sizes (34 and 42) ceramic magnets are used to minimize cost. The armature in the foreground shows the skewed slotting used to ensure smooth torque production. Note: The size number refers to the body diameter in tenths of an inch; a size 15 motor has a diameter of 1.5 inches (38.1 mm) (Photograph by courtesy of Evershed and Vignoles Ltd)*

ductors) on the rotor move, the magnetic part (either iron or permanent magnet) remaining stationary. This principle is adopted in 'ironless rotor' and 'printed armature' motors.

In the ironless rotor or moving-coil type, the armature conductors are formed as a thin-walled cylinder, sometimes referred to as a basket winding by analogy with a cane wastepaper basket. This cylinder, consisting essentially of nothing more than varnished wires wound in skewed form together with the disc-type commutator, is the only part of the motor to rotate. Inside the armature sits a cylindrical

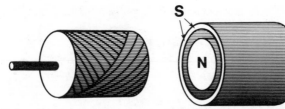

Figure 4.12   *Ironless rotor d.c. motor. The commutator (not shown) is usually of the disc type*

permanent magnet which provides the radial flux, and outside it is a steel cylindrical shell which completes the magnetic circuit, as shown in Figure 4.12

Needless to say the absence of any slots to support the armature winding results in a relatively fragile structure, which is therefore limited to diameters of not much over 1 cm. Because of their small size they are often known as micromotors, and are very widely used in cameras, video systems, card readers etc.

The printed armature type is altogether more robust, and is made in sizes up to a few kW. They are generally made in disc or pancake form, with the direction of flux axial and the armature current radial. The armature conductors resemble spokes on a wheel, the conductors themselves being formed on a lightweight disc. Early versions were made by using printed-circuit techniques, but pressed fabrication is now more common. Since there are usually at least a hundred armature conductors, the torque remains almost constant as the rotor turns, which allows them to produce very smooth rotation at low speed. Inertia and armature inductance are very low, giving a good dynamic response, and the short and fat shape makes them suitable for applications such as machine tools and disc drives where axial space is at a premium.

Position control

As mentioned earlier many servo motors are used in closed-loop postion control applications, so it is appropriate to

look briefly at how this is achieved. Later (in Chapter 8) we
will see that the stepping motor provides a simpler open-
loop alternative method of position control.

In the example shown in Figure 4.13, the angular position
of the output shaft is intended to follow the reference voltage (θ_{ref}), but it should be clear that if the motor drives a
toothed belt linear outputs can also be obtained. The
potentiometer mounted on the output shaft provides a feedback voltage proportional to the actual position of the output shaft. The voltage from this potentiometer must be a
linear function of angle, and must not vary with temperature, otherwise the accuracy of the system will be in
doubt.

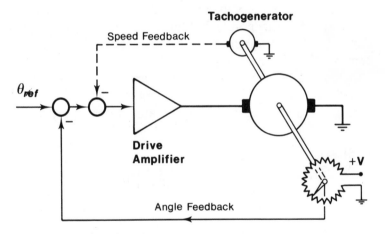

Figure 4.13   *Closed-loop angular position control using d.c. motor and angle
feedback from a servo-type potentiometer*

The feedback voltage (representing the actual angle of the
shaft) is subtracted from the reference voltage (representing
the desired position) and the resulting position error signal is
amplified and used to drive the motor so as to rotate the
output shaft in the desired direction. When the output shaft
reaches the target position, the position error becomes zero,
no voltage is applied to the motor, and the output shaft

remains at rest. Any attempt to physically move the output shaft from its target position immediately creates a position error and a restoring torque is applied by the motor.

The dynamic performance of the simple scheme described above is very unsatisfactory as it stands. In order to achieve a fast response and to minimize position errors caused by static friction, the gain of the amplifier needs to be high, but this in turn leads to a highly oscillatory response which is usually unacceptable. For some fixed-load applications matters can be improved by adding a compensating network in series with the amplifier, but the best solution is to use 'tacho' feedback (shown dotted in Figure 4.13) in addition to the main position feedback loop.

Tacho feedback clearly has no effect on the static behaviour (since the voltage from the tacho is proportional to the speed of the motor), but has the effect of increasing the damping of the transient response. The gain of the amplifier can therefore be made high in order to give a fast response, and the degree of tacho feedback can then be adjusted to provide the required damping (see Figure 4.14). Many servo motors have an integral tachogenerator for this purpose.

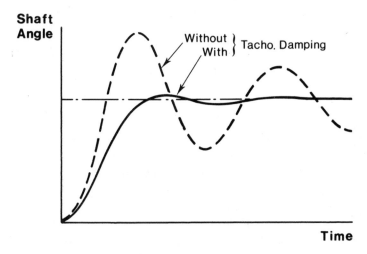

Figure 4.14   *Typical step responses for closed-loop position control system, showing the improved damping obtained by the addition of tacho feedback*

The example above dealt with an analogue scheme in the interests of simplicity, but digital position control schemes are now gradually taking precedence, especially when brushless motors (see Chapter 9) are used. Complete 'controllers on a card' are available as off-the-shelf items, and these offer ease of interface to other sytems as well as providing for improved flexibility in shaping the dynamic response.

5

INDUCTION MOTORS – ROTATING FIELD, SLIP AND TORQUE

INTRODUCTION

Judged in terms of fitness for purpose coupled with simplicity, the induction motor must rank alongside the screw-thread as one of mankind's best inventions. It is not only supremely elegant as an electromechanical energy converter, but is also by far the most important, with something like one third of all the electricity generated being converted back to mechanical energy in induction motors. Despite playing a key role in modern industrial society, it remains largely unnoticed because of its workaday role in unglamorous industrial surroundings driving machinery, pumps, fans, compressors, conveyors, hoists, and a host of other routine but vital tasks. It will doubtless continue to dominate these fixed-speed applications, but, thanks to the availability of reliable variable-frequency inverters, it is also establishing itself in the controlled-speed arena.

Like the d.c. motor, the induction motor develops torque by the interaction of axial currents on the rotor and a radial magnetic field produced by the stator. But whereas in the d.c. motor the 'work' current has to be fed into the rotor by means of brushes and a commutator, the currents in the rotor of the induction motor are induced by electromagnetic action, hence the name 'induction' motor. The stator winding therefore not only produces the magnetic field (the 'excitation'), but also supplies the energy which is converted to mechanical output.

Other differences between the induction motor and the d.c. motor are firstly that the supply to the induction motor is a.c. (usually three-phase, but in smaller sizes single-phase); secondly that the magnetic field in the induction motor rotates relative to the stator, while in the d.c. motor it is stationary; and thirdly that both stator and rotor in the induction motor are non-salient (i.e. effectively smooth) whereas the d.c. motor stator has projecting poles or saliencies which define the position of the field windings.

Given these differences we might expect to find major contrasts between the performance of the two types of motor, and it is true that their inherent characteristics exhibit distinctive features. But there are also many aspects of behaviour which are similar, as we shall see. Perhaps most important from the user's point of view is that there is no dramatic difference in size or weight between an induction motor and a d.c. motor giving the same power at the same base speed, though the induction motor will almost always be much cheaper. The similarity in size is a reflection of the fact that both types employ similar amounts of copper and iron, while the difference in price stems from the much simpler construction of the induction motor.

Outline of approach

To understand how an induction motor operates, we must first unravel the mysteries of the rotating magnetic field. We shall see later that the rotor is effectively dragged along by the rotating field, but that it can never run quite as fast as the field. When we want to control the speed of the rotor, the best way will be to control the speed of the field.

Our look at the mechanism of the rotating field will focus on the stator windings because they act as the source of the flux. In this part of the discussion we will ignore the presence of the rotor coils. This makes it much easier to understand what governs the speed of rotation and the magnitude of the field, which are the two factors that most influence motor behaviour.

Having established how the rotating field is set up, and what its speed and strength depend on, we move on to examine the rotor, concentrating on how it behaves when exposed to the rotating field, and discovering how the torque varies with speed. In this section we assume – again for the sake of simplicity – that the rotating flux set up by the stator is not influenced by the rotor.

Finally we turn attention to the interaction between the rotor and stator, verifying that our earlier assumptions are well justified. Having done this we are in a position to examine the 'external characteristics' of the motor, i.e. the variation of motor torque and current with speed. These are the most important characteristics from the point of view of the user.

In discussing how the motor operates the approach leans heavily on first building up a picture of the main air-gap flux. All of the main characteristics which are of interest to the user can be explained and understood once a clear idea has been formed of what the flux wave is, what determines its amplitude and speed, and how it interacts with the rotor to produce torque.

The amount of mathematics used has been minimised, and all but the simplest equivalent circuit have been eschewed in favour of a physical explanation. Most conventional treatments make extensive use of equivalent circuits, but experience indicates that whilst they can be very illuminating in expert hands, it is seldom possible for the non-specialist to use them profitably.

THE ROTATING MAGNETIC FIELD

Before we look at how the rotating field is produced, we should be clear what it actually is. Because both the rotor and stator iron surfaces are smooth (apart from the regular slotting), and are separated by a small air-gap, the flux produced by the stator windings crosses the air-gap radially. The behaviour of the motor is to a large extent dictated by this radial flux, so we will concentrate first on establishing a mental picture of what is meant by the 'flux wave' in an induction motor.

The pattern of flux in an ideal 4-pole motor supplied from a balanced three-phase source is shown in Figure 5.1(a), at three different instants of time. The term '4-pole' reflects the fact that flux leaves the stator from two N poles, and returns at two S poles. Note however that there are no physical features of the stator iron which mark it out as being 4-pole, rather than say 2-pole or 6-pole. As we shall see, it is the layout of the stator coils which sets the pole number.

If we plot the variation of the air-gap flux density with respect to distance round the stator, at each of the three instants of time, we get the patterns shown in Figure 5.1(b). The first feature to note is that the radial flux density varies sinusoidally in space. There are two N peaks and two S peaks, but the transition from N to S occurs in a smooth sinusoidal way, giving rise to the term 'flux wave'. The distance from the centre of one N pole to the centre of the adjacent S pole is called the pole-pitch, for obvious reasons.

Returning to Figure 5.1(b), we note that after one quarter of a cycle of the mains frequency, the flux wave retains its original shape, but has moved round the stator by half a pole-pitch, while after half a cycle it has moved round by a full pole-pitch. If we had plotted the patterns at intermediate times, we would have discovered that the wave maintained a constant shape, and progressed smoothly, advancing at a uniform rate of two pole-pitches per cycle of the mains. The term 'travelling flux wave' is thus an appropriate one to describe the air-gap field.

For the 4-pole wave here, one complete revolution takes two cycles of the supply, so the speed is 25 revs/sec (1 500 rev/min) with a 50 Hz supply, or 30 rev/sec (1 800 rev/min) at 60 Hz. The general expression for the speed of the field (which is known as the synchronous speed) N_s, in rev/min is

$$N_s = \frac{120f}{p}. \tag{5.1}$$

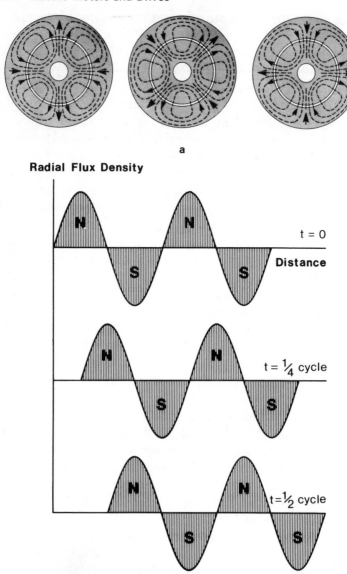

b

Figure 5.1   (a) *Flux pattern in a 4-pole induction motor at three successive instants of time, each one-quarter of a cycle apart,* (b) *radial flux density distribution in the air-gap at the three instants shown in Figure 5.1a* where p is the pole-number. The pole-number must be an even integer, since for every N pole there must be a S pole. Synchronous speeds for commonly-used pole-numbers are given in the table below.

Pole Number	50 Hz	60 Hz
2	3000	3600
4	1500	1800
6	1000	1200
8	750	900
10	600	720
12	500	600

Synchronous Speeds, in Revs/Min

We can see from the table that if we want the field to rotate at intermediate speeds, we will have to be able to vary the supply frequency, and this is what happens in inverter-fed motors, which are dealt with in Chapter 7.

Production of rotating magnetic field

Now that we have a picture of the field, we turn to how it is produced. If we inspect the stator winding of an induction motor we find that it consists of a uniform array of identical coils, located in slots. The coils are in fact connected to form three identical groups or phases, distributed around the stator, and symmetrically displaced with respect to one another. The three phase-windings are connected either in star (wye) or delta, as shown in Figure 5.2.

The three phase-windings are connected directly to a three-phase a.c. supply, and so the currents (which set up the flux) are of equal amplitude but differ in time-phase by one third of a cycle (120°), forming a balanced three-phase set.

Plate 5.1   *Stator of three-phase TEFV induction motor. The semi-closed slots of the stator core obscure the active sides of the stator coils, but the ends of the coils are just visible beneath the binding tape (Photograph by courtesy of GEC Electromotors Ltd)*

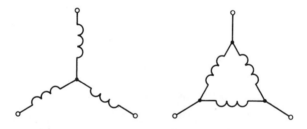

Figure 5.2   *Star (Wye) and Delta connection of the three phase-windings of a 3-phase induction motor*

Field produced by each phase-winding

The aim of the winding designer is to arrange the layout of the coils so that each phase-winding, acting alone, produces an MMF wave (and hence an air-gap flux wave) of the desired pole-number, and with a sinusoidal variation of

Plate 5.2   *Cage rotor for induction motor. The rotor conductor bars and end rings are cast in aluminium, and the blades attached to the end rings serve as a fan for circulating internal air. An external fan will be mounted on the non-drive end (Photograph by courtesy of GEC Electromotors Ltd)*

amplitude with angle. Getting the right pole-number is not difficult: we simply have to choose the right number and pitch of coils, as shown in the developed diagrams of an elementary 4-pole winding shown in Figure 5.3.

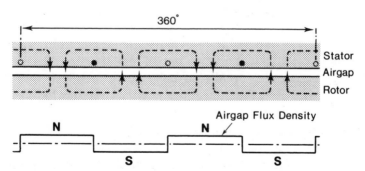

Figure 5.3   *Developed diagram showing elementary 4-pole, single-layer stator winding consisting of 4 conductors spaced by 90°. The 'go' side of each conductor (shown ○) carries positive current at the instant shown, while the 'return' side (shown ●) carries negative current*

From Figure 5.3 we see that although the arrangement shown gives us the correct number of poles (i.e. 4), the field pattern is rectangular, whereas we want it to be sinusoidal. We can improve matters by adding more coils in the adjacent slots, as shown in Figure 5.4. All the coils have the same number of turns, and carry the same current. The addition of the extra slightly-displaced coils gives rise to the stepped waveform of MMF and air-gap flux density shown in Figure 5.4. It is still not sinusoidal, but is better than the original rectangle.

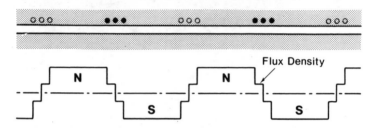

Figure 5.4　*Developed diagram showing flux density produced by one phase of a single-layer winding having three slots per pole per phase*

It turns out that if we were to insist on having a perfect sinusoidal flux, we would have to distribute the coils of one phase sinusoidally over the whole periphery of the stator. This is not a practicable proposition, firstly because we would have to vary the number of turns per coil from point to point, and secondly because we want the coils to be in slots, so it is impossible to avoid some measure of discretisation in the layout. For economy of manufacture we are also obliged to settle for all the coils being identical, and we must make sure that the three identical phase-windings fit together in such a way that all the slots are fully utilised.

Despite these constraints we can get remarkably close to the ideal sinusoidal pattern, especially when we use a 'two-layer' winding. A typical arrangement of one phase is shown in Figure 5.5.

This type of winding is almost universal in all but small induction motors, the coils in each phase being grouped to-

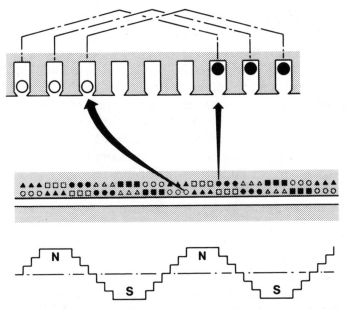

Figure 5.5   *Developed diagram showing layout of windings in a 3-phase,
4-pole, two-layer induction motor winding, together with the flux density
produced by one phase acting alone. (The three phases are represented by
circle, triangle and square, the go and return sides of the coils being shown by
outline and solid symbols respectively. The upper detail shows how the
two-layer coils are arranged.)*

gether to form 'phase-bands' or 'phase-belts'. We see from
Figure 5.5 that the field produced by one phase (A is shown)
is a fair approximation to a sinusoid.

The current in each phase pulsates at the supply frequency, so the field produced by phase A pulsates in sympathy with the current in phase A, the axis of each pole
remaining fixed in space, but its polarity changing from N to
S and back once per cycle. There is no hint of any rotation in
the field of one phase, but when the three phases are
combined, matters change dramatically.

Resultant field

We can see in Figure 5.5 that the windings of phases B and C
are identical with that of phase A apart from the fact that

they are displaced in space by one third and two thirds of a pole-pitch respectively. Phases B and C therefore also produce pulsating fields, along their own fixed axes in space. But the currents in phases B and C also differ in time-phase from the current in phase A, lagging by one third and two thirds of a cycle respectively. To find the resultant field we must therefore superimpose the fields of the three phases, taking account of the spatial differences between windings, and the time differences between the currents.

When we plot the resultant field for the complete 4-pole winding shown in Figure 5.5, for several discrete times during one complete cycle, we obtain the patterns shown in Figure 5.6.

We see that the three pulsating fields combine beautifully and lead to a resultant 4-pole field which rotates at a uniform rate, advancing by two pole-pitches for every cycle of the mains. The resultant field is not exactly sinusoidal in shape (though it is actually more sinusoidal than the field produced by the individual phase-windings), and its shape varies a little from instant to instant; but these are minor worries. The resultant field is amazingly close to the ideal travelling wave and yet the winding layout is simple and easy to manufacture. This is an elegant engineering achievement, however one looks at it.

Direction of rotation

The direction of rotation depends on the order in which the currents reach their maxima, i.e. on the phase-sequence of the supply. Reversal of direction is therefore simply a matter of interchanging any two of the lines connecting the windings to the supply.

Main (air-gap) flux and leakage flux

Broadly speaking the motor designer shapes the stator and rotor teeth to encourage as much as possible of the flux produced by the stator to pass right down the rotor teeth, so

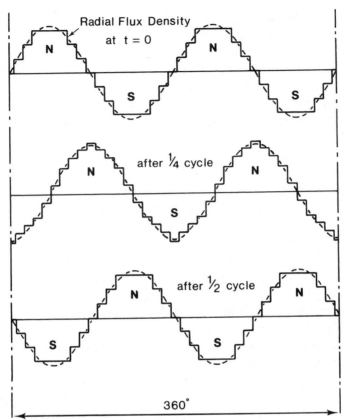

Figure 5.6   *Resultant air-gap flux density produced by the complete 3-phase winding*

that before completing its path back to the stator it is fully linked with the rotor conductors which are located in the rotor slots. We will see later that this tight magnetic coupling between stator and rotor windings is necessary for good running performance, and the field which provides the coupling is of course the main or air-gap field, which we are in the midst of discussing.

In practice the vast majority of the flux produced by the stator is indeed main or 'mutual' flux. But there is some flux which bypasses the rotor conductors, linking only with the stator winding, and known as stator leakage flux. Similarly not all the flux produced by the rotor currents (see later) links the stator, but some the rotor leakage flux links only the rotor conductors.

On the face of it these leakage fluxes would seem to be unwelcome imperfections, which we should go out of our way to minimize. However whilst the majority of aspects of performance are certainly enhanced if the leakage is as small as possible, others (notably the large current drawn from the mains when the motor is started from rest) are made much worse if the coupling is too good. So we have the somewhat paradoxical situation in which the designer finds it comparatively easy to lay out the windings to produce a good main flux, but is then obliged to juggle the detailed design of the slots in order to obtain just the right amount of leakage flux to give acceptable all-round performance.

The weight which attaches to the matter of leakage flux is reflected in the prominent part played by leakage reactance in equivalent circuit models of the induction motor. This is helpful for designers, who want to assess the effects of changing the leakage, but of little use to the user. We shall therefore only make occasional references to leakage reactance, and then only in well-defined contexts. In general, where the term 'flux' is used, it refers to the main air-gap field.

Magnitude of rotating flux wave

We have already seen that the speed of the flux wave is set by the pole number of the winding and the frequency of the supply. But what is it that determines the amplitude of the field?

To answer this question we can continue to neglect the fact that under normal conditions there will be induced currents in the rotor. We might even find it easier to imagine that the rotor conductors have been removed altogether: this may seem a drastic assumption, but will prove justified later. The stator windings are assumed to be connected to a balanced three-phase a.c. supply so that a balanced set of currents flows in the windings. We denote the phase voltage by V, and the current in each phase by I_m, where the subscript m denotes 'magnetizing' or flux-producing current.

From the discussion in Chapter 1 we know that the magnitude of the flux wave (B_m) is proportional to the winding MMF, and is thus proportional to I_m. But what we really want to know is how the flux density depends on the supply voltage and frequency, since these are the only two parameters over which we have control.

To guide us to the answer, we must first ask what effect the travelling flux wave will have on the stator winding. Every stator conductor will of course be cut by the rotating flux wave, and will therefore have an e.m.f. induced in it. Since the flux wave varies sinusoidally in space, and cuts each conductor at a constant velocity, a sinusoidal e.m.f is induced in each conductor. The magnitude of the e.m.f. is proportional to the magnitude of the flux wave (B_m), and to the speed of the wave (i.e. to the supply frequency f). The frequency of the induced e.m.f. depends on the time taken for one N pole and one S pole to cut the conductor. We have already seen that the higher the pole-number, the slower the field rotates, but we found that the field always advances by two pole-pitches for every cycle of the mains. The frequency of the e.m.f. induced in the stator conductors is therefore the same as the supply frequency, regardless of the pole-number.

The e.m.f in each complete phase winding (E) is the sum of the e.m.f.s in the phase coils, and will thus also be at supply frequency. (The alert reader will realize that whilst the e.m.f in each coil has the same magnitude, it will differ in time phase, depending on the geometrical position of the coil. Most of the coils in each phase-band are close together, however, so their e.m.f.s – though slightly out of phase – will more or less add up directly.)

If we were to compare the e.m.f.s in the three complete phase windings, we would find that they were of equal amplitude, but out of phase by one third of a cycle (120°), thereby forming a balanced three-phase set. This result could have been anticipated from the overall symmetry. It means that we can consider one of the phases only in the rest of the discussion.

So we find that when an alternating voltage V is applied, an alternating e.m.f., E, is induced. We can represent this state of affairs by the a.c. equivalent circuit shown in Figure 5.7.

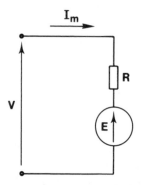

Figure 5.7   *Simple equivalent circuit for the induction motor under no-load conditions*

The resistance shown in Figure 5.7 is the resistance of one complete phase-winding. Note that the e.m.f. E is shown as opposing the applied voltage V. This must be so, otherwise we would have a runaway situation in which the voltage V produced the current I_m which in turn set up an e.m.f. E, which added to V, which further increased I_m and so on *ad infinitum*.

Applying Kirchoff's law to the a.c. circuit in Figure 5.7 yields

$$V = I_m R + E \qquad (5.2)$$

We find in practice that the term $I_m R$ (which represents the volt drop due to winding resistance) is usually very much less than the applied voltage V. In other words most of the applied voltage is accounted for by the opposing e.m.f., E. Hence we can make the approximation

$$V \approx E \qquad (5.3)$$

But we have already seen that the e.m.f. is proportional to B_m and to f, i.e.

$$E \propto B_m f. \qquad (5.4)$$

So by combining equations 5.3 and 5.4 we obtain

$$B_m = k\frac{V}{f}. \qquad (5.5)$$

Equation 5.5 is of fundamental importance in induction motor operation. It shows that if the supply frequency is constant, the flux in the air-gap is directly proportional to the applied voltage, or in other words the voltage sets the flux. We can also see that if we raise or lower the frequency (in order to increase or reduce the speed of rotation), we will have to raise or lower the voltage in proportion if, as is usually the case, we want the magnitude of the flux to remain constant.

It may seem a paradox that having originally homed-in on the magnetizing current I_m as being the source of the MMF which in turn produces the flux, we find that the actual value of the flux is governed only by the applied voltage and frequency, and I_m does not appear at all in equation 5.5.

We can perhaps see why this is so by looking again at Figure 5.7 and asking what would happen if, for some reason, the e.m.f. E were to reduce. We would find that I_m would increase, which in turn would lead to a higher MMF, more flux, and hence to an increase in E. There is clearly a negative feedback effect taking place, which continually tries to keep E equal to V. It is rather like the d.c. motor (Chapter 3) where the speed of the unloaded motor always adjusted itself so that the back e.m.f. equalled the applied voltage. Here, the magnetizing current always adjusts itself so that the induced e.m.f. is almost equal to the applied voltage.

Needless to say this does not mean that the magnetizing current is arbitrary, but to calculate it we would have to know the number of turns in the winding, the length of the air-gap (from which we could calculate the gap reluctance) and the reluctance of the iron paths. From a user point of view there is no need to delve further in this direction. We should however recognize that the reluctance will be dominated by the air-gap, and that the magnitude of the magnetizing current will therefore depend mainly on the size of the gap. The larger the gap, the bigger the magnetizing current. Since the magnetizing current contributes to stator copper loss, but not to useful output power, we would like it to be a small as possible, so we find that induction motors usually have the smallest air-gap which is consistent with providing the necessary mechanical clearances. Despite the small air-gap the magnetizing current can be appreciable: in a 4-pole motor, it may be typically 50 per cent of the full-load current, and even higher in 6-pole and 8-pole designs.

Excitation power and VA

The setting up of the travelling wave by the magnetizing current amounts to the provision of 'excitation' for the motor. Some energy is stored in the magnetic field, but since the amplitude remains constant once the field has been established, no power input is needed to sustain the field. We therefore find that under the conditions discussed so far, i.e. in the absence of any rotor currents, the power input to the motor is very small. (We should perhaps note that the rotor currents in a real motor are very small when it is running light, so the hypothetical situation we are looking at is not so far removed from reality as we may have supposed.)

Ideally the only source of power losses would be the copper losses in the stator windings, but to this must be added the 'iron losses' which arise from eddy currents and hysteresis in the laminated cores of rotor and stator. However we have seen that the magnetizing current can be quite large, its value being largely determined by the air-gap, so we can expect an appreciable current to be drawn from the supply, but very little real power. The VA will therefore be appreciable, but the power-factor will be very low, the magnetizing current lagging the supply voltage by almost 90°, as shown in the phasor diagram (Figure 5.8).

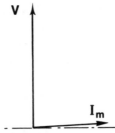

Figure 5.8   *Phasor diagram under no-load conditions, showing magnetizing current I_m*

Viewed from the supply the stator looks more or less like a pure inductance, a fact which we would expect intuitively given that – having ignored the rotor circuit – we are left with only an arrrangement of flux-producing coils surrounded by a good magnetic circuit.

Summary

When the stator is connected to a three-phase supply, a rotating magnetic field is set up in the air-gap. The speed of rotation of the field is directly proportional to the frequency of the supply, and inversely proportional to the pole-number of the winding. The magnitude of the flux wave is proportional to the applied voltage, and inversely proportional to the frequency.

When the rotor circuits are ignored, the real power drawn from the mains is small, but the magnetizing current itself can be quite large, giving rise to a significant reactive power demand from the mains.

TORQUE PRODUCTION

In this section we begin with a brief description of rotor types, and introduce the notion of 'slip', before moving on to explore how the torque is produced, and investigate the variation of torque with speed. We will find that the behaviour of the rotor varies widely according to the slip, and we therefore look separately at low and high values of slip.

Throughout this section we will assume that the rotating magnetic field is unaffected by anything which happens on the rotor side of the air-gap. Later, we will see that this assumption is pretty well justified.

Rotor construction

Two types of rotor are used in induction motors. In both the rotor 'iron' consists of a stack of steel laminations with evenly-spaced slots punched around the circumference.

The cage rotor is by far the most common: each rotor slot contains a solid conductor bar and all the conductors are physically and electrically joined together at each end of the rotor by conducting end-rings (Figure 5.9). The conductors may be of copper, in which case the end-rings are brazed-on. Or, in small and medium sizes, the rotor conductors and end rings can be die-cast in aluminium.

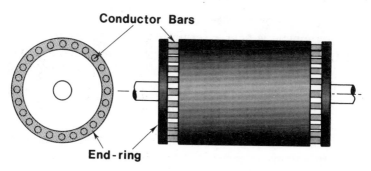

Figure 5.9   *Cage rotor construction*

The term squirrel cage was widely used at one time and the origin should be clear from Figure 5.9. The rotor bars and end-rings are reminiscent of the rotating cages used in bygone days to exercise small rodents (or rather to amuse their human captors).

The absence of any means for making direct electrical connection to the rotor underlines the fact that in the induction motor the rotor currents are induced by the air-gap field. It is equally clear that because the rotor cage comprises permanently short-circuited conductor bars, no external control can be exercised over the resistance of the rotor circuit once the rotor has been made. This is a significant drawback which can be avoided in the second type of rotor, which is known as the 'wound-rotor' or 'slipring' type.

In the wound rotor, the slots accommodate a set of three phase-windings very much like those on the stator. The windings are connected in star, with the three ends brought out to three slip-rings (Figure 5.10). The rotor circuit is thus open, and connection can be made via brushes bearing on the sliprings. In particular, the resistance of the rotor circuit can be increased by adding external resistance, as indicated in Figure 5.10. Adding resistance in appropriate circumstances can be beneficial, as we shall see.

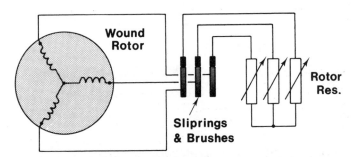

Figure 5.10   *Wound rotor, showing sliprings and brushes to provide connection to the external (stationary) 3-phase resistance*

Cage-rotors are usually cheaper to manufacture, and are very robust and reliable. Until the advent of variable-frequency inverter supplies, however, the superior control which was possible from the slip-ring type meant that the extra expense of the wound rotor and its associated control gear were frequently justified, especially for high-power machines. Nowadays comparatively few are made, and those that are are invariably large. But many old motors remain in service, so they are included in the discussion.

Slip

A moment's thought will show that the behaviour of the rotor depends very much on its relative velocity with respect to the rotating field. If the rotor is stationary, for example, the rotating field will cut the rotor conductors at synchronous speed, thereby inducing a high e.m.f. in them. On the other hand, if the rotor was running at the synchronous speed, its relative velocity with respect to the field would be zero, and no e.m.f.s would be induced in the rotor conductors.

The relative velocity between the rotor and the field is known as the slip. If the speed of the rotor is N, the slip speed is $N_s - N$, where N_s is the synchronous speed of the field, usually expressed in rev/min. The slip (as distinct from slip speed) is the normalized quantity defined by

$$s = \frac{N_s - N}{N_s}. \qquad (5.6)$$

and is usually expressed either as a ratio as in equation 5.6, or as a percentage. A slip of 0 therefore indicates that the rotor speed is equal to the synchronous speed, while a slip of 1 corresponds to zero speed or 'locked-rotor' conditions.

Rotor induced e.m.f., current and torque

The rate at which the rotor conductors are cut by the flux – and hence their induced e.m.f. – is directly proportional to the slip, with no induced e.m.f. at synchronous speed (s = 0) and maximum induced e.m.f. when the rotor is stationary (s = 1).

The frequency of rotor e.m.f. is also directly proportional to slip, since the rotor effectively slides with respect to the flux-wave, and the higher the relative speed, the more times in a second each rotor conductor is cut by a N and a S pole. At synchronous speed the frequency is zero, while at standstill, the rotor frequency is equal to the supply frequency. These relationships are shown in Figure 5.11.

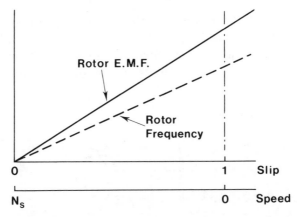

Figure 5.11   *Variation of rotor induced e.m.f and frequency with speed and slip*

Although the e.m.f. induced in every rotor bar will have the same magnitude and frequency, they will not be in phase. At any particular instant, bars under the peak N poles of the field will have maximum positive voltage in them, those under the peak S poles will have maximum negative voltage, (i.e. 180° phase shift) and those in between will have varying degrees of phase shift. The pattern of instantaneous voltages in the rotor is a replica of the flux wave, and the voltage wave therefore moves relative to the rotor at slip speed, as shown in Figure 5.12.

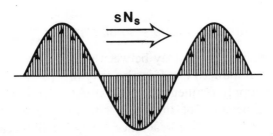

Figure 5.12   *Pattern of induced e.m.f.s in rotor conductors. The rotor 'voltage wave' moves at a speed of sN_s with respect to the rotor surface*

Since all the rotor bars are short-circuited by the end-rings, the induced voltages will drive currents along the rotor bars, the currents forming closed paths through the end-rings, as shown in the developed diagram (Figure 5.13).

End-ring

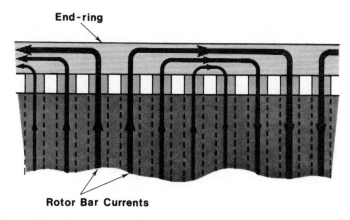

Rotor Bar Currents

Figure 5.13 *Instantaneous sinusoidal pattern of rotor currents in rotor bars and end-rings. Only one pole-pitch is shown, but the pattern is repeated*

The axial currents in the rotor bars will interact with the radial flux wave to produce the driving torque of the motor, which will act in the same direction as the rotating field, the rotor being dragged along by the field. We note that slip is essential to this mechanism, so that it is never possible for the rotor to catch up with the field, as there would then be no rotor e.m.f., no current, and no torque.

Rotor currents and torque – small slip

When the slip is small (say between 0 and 10 per cent), the frequency of induced e.m.f. is also very low (between 0 and 5 Hz if the supply frequency is 50 Hz). At these low frequencies the impedance of the rotor circuits is predominantly resistive, the inductive reactance being small because the rotor frequency is low.

The current in each rotor conductor is therefore in time-phase with the e.m.f. in that conductor, and the rotor current-wave is therefore in space-phase with the rotor e.m.f. wave, which in turn is in space-phase with the flux wave. This situation is shown in Figure 5.14.

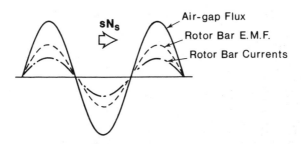

Figure 5.14   *Pattern of air-gap flux, and induced e.m.f. and current in cage rotor bars at low values of slip*

To calculate the torque we first need to evaluate the 'BI$_r$l' product in order to obtain the tangential force on each rotor conductor. The torque is then given by the total force multiplied by the rotor radius. We can see from Figure 5.14 that where the flux has a positive peak, so does the rotor current, so that particular bar will contribute a high tangential force to the total torque. Similarly, where the flux has its maximum negative peak, the induced current is maximum and negative, so the tangential force is again positive. We don't need to work out the torque in detail, but it is clear that the resultant will be given by an equation of the form

$$T = kBI_r \qquad (5.7)$$

where B and I$_r$ denote the amplitudes of the flux and rotor current waves respectively. Provided that there are a large number of rotor bars (which is a safe bet in practice), the waves shown in Figure 5.14 will remain the same at all instants of time, so the torque remains constant as the rotor rotates.

If the supply voltage and frequency are constant, the flux will be constant (See equation 5.5). The rotor e.m.f. (and hence I_r) is then proportional to slip, so we can see from equation 5.7 that the torque is directly proportional to slip. We must remember that this discussion relates to low values of slip only, but since this is the normal running condition, it is extremely important.

The torque-speed relationship for small slips is thus a straight-line, as shown in Figure 5.15.

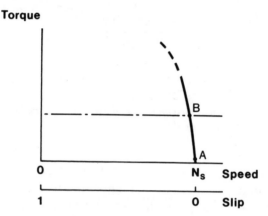

Figure 5.15   *Torque-speed relationship for low values of slip*

If the motor is unloaded, it will need very little torque to keep running – only enough to overcome friction in fact – so an unloaded motor will run with a very small slip at just below the synchronous speed, as shown at A in Figure 5.15.

When the load is increased, the rotor slows down, and the slip increases, thereby inducing more rotor e.m.f. and current, and thus more torque. The speed will settle when the slip has increased to the point where the developed torque equals the load torque – e.g. point B in Figure 5.15.

Induction motors are usually designed so that their full-load torque is developed for small values of slip. Small ones typically have a full-load slip of 8 per cent, large ones around 1 per cent. At the full-load slip, the rotor conductors will be carrying their safe maximum continuous current, and if the slip is any higher, the rotor will begin to overheat. This overload region is shown by the dotted line in Figure 5.15.

The torque-slip (or torque-speed) characteristic shown in Figure 5.15 is a good one for most applications, because the speed only falls a little when the load is raised from zero to its full value. We note that, in this normal operating region, the torque-speed curve is very similar to that of a d.c. motor, which explains why both d.c. and induction motors are often in contention for constant-speed applications.

Rotor currents and torque – large slip

As the slip increases, the rotor e.m.f. and rotor frequency both increase in direct proportion to the slip. At the same time the rotor leakage reactance, which was negligible at low slip (low rotor frequency) begins to be appreciable in comparison with the rotor resistance. Hence although the induced current continues to increase with slip, it does so more slowly than at low values of slip, as shown in Figure 5.16.

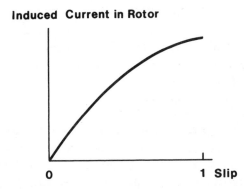

Figure 5.16   *Magnitude of current induced in rotor over the full range of slip*

At high values of slip, the rotor current also lags behind the rotor e.m.f. because of the inductive reactance. The alternating current in each bar reaches its peak well after the induced voltage, and this in turn means that the rotor current-wave has a space-lag with respect to the rotor e.m.f. wave (which is in space-phase with the flux wave). This space-lag is shown by the angle ϕ_r in Figure 5.17.

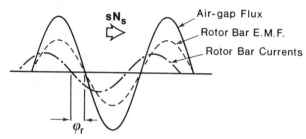

Figure 5.17   *Pattern of air-gap flux, and induced e.m.f. and current in cage rotor bars at high values of slip (c.f. Figure 5.14)*

The space-lag means that the peak flux and peak rotor currents no longer coincide, which is bad news from the point of view of torque production, because whilst we have high values of both flux density and current, they do not occur simultaneously at any point around the periphery. What is worse is that at some points we even have flux density and currents of opposite sign, so over those regions of the rotor surface the torque contributed will actually be negative. The overall torque will still be positive, but is much less than it would be if the flux and current waves were in phase. We can allow for the unwelcome space-lag by modifying equation 5.7, to obtain a more general expression for torque as

$$T = kBI_r \cos \phi_r. \tag{5.8}$$

Equation 5.7 is merely a special case of equation 5.8, but only applies under low-slip conditions where $\phi_r \propto 1$.

It turns out that for most cage rotors the term $\cos \phi_r$ reduces more quickly than the current (I_r) increases, so that at some slip between 0 and 1 the developed torque reaches its maximum value. This is shown in the complete torque-speed characteristic in Figure 5.18, the peak torque actually occuring for the value of slip at which the rotor reactance is equal to the rotor resistance.

We will return to the torque-speed curve after we have checked that when we allow for the interaction of the rotor with the stator, our interim conclusions regarding torque production remain valid.

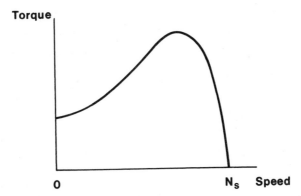

Figure 5.18   *Typical complete torque-speed characteristic for cage induction motor*

INFLUENCE OF ROTOR CURRENT ON FLUX

Up to now all our discussion has been based on the assumption that the rotating magnetic field remains constant, regardless of what happens on the rotor. We have seen how torque is developed, and that mechanical output power is produced. We have focused attention on the rotor, but the output power must be provided from the stator winding, so we must turn attention to the behaviour of the whole motor, rather than just the rotor. Several questions spring to mind.

Firstly, what happens to the rotating magnetic field when the motor is working? Won't the MMF of the rotor currents cause it to change? Secondly, how does the stator know when to start supplying real power across the air-gap to allow the rotor to do useful mechanical work? And finally, how will the currents drawn by the stator vary as the slip is changed?

These are demanding questions, for which full treatment is beyond our scope. But we can deal with the essence of the matter without too much difficulty.

Reduction of flux by rotor current

We should begin by recalling that we have already noted that when the rotor currents are negligible (s = 0), the e.m.f. which the rotating field induces in the stator winding is very nearly equal to the applied voltage. A reactive current (which we termed the magnetizing current) flows into the windings, to set up the rotating flux. Any slight tendency for the flux to fall is immediately detected by a corresponding slight reduction in e.m.f. which is reflected in a disproportionately large increase in magnetizing current, which thus opposes the tendency for the flux to fall.

Exactly the same mechanism comes into play when the slip increases from zero, and rotor currents are induced. The rotor current wave gives rise to a rotor MMF wave, which rotates at slip speed (s N_s) relative to the rotor. But the rotor is rotating at a speed of $(1 - s)N_s$, so that when viewed from the stator, the rotor MMF wave rotates at synchronous speed. The rotor MMF wave would, if unchecked, cause its own 'rotor flux wave', rotating at synchronous speed in the air-gap, in much the same way that the stator magnetizing current originally set up the flux wave. The rotor flux wave would oppose the original flux wave, causing the resultant flux wave to reduce.

However, as soon as the resultant flux begins to fall, the stator e.m.f. reduces, thereby admitting more current to the stator winding, and increasing its MMF. A very small drop in e.m.f. in the stator is sufficient to cause a large increase in the current drawn from the mains. The 'extra' stator MMF effectively 'cancels' the MMF produced by the rotor currents, leaving the resultant MMF (and hence the rotating flux wave) virtually unchanged.

There **must** be a **small** drop in the flux of course, to alert the stator to the presence of rotor currents. But because of the delicate balance between the applied voltage and the induced e.m.f. in the stator the change in flux with load is very small, at least over the normal operating speed-range, where the slip is small. In large motors, the drop in flux over the normal operating region is typically less than 1 per cent, rising to perhaps 10 per cent in a small motor. As far as our conclusions regarding torque are concerned, we see that our original assumption that the flux was constant is near enough correct when the slip is small. We will find it helpful and convenient to continue to treat the flux as constant (for given stator voltage and frequency) when we turn later to methods of controlling the normal running speed.

It has to be admitted, however, that at high values of slip (i.e. low rotor speeds), we cannot expect the main flux to remain constant, and in fact we would find in practice that when the motor was first switched on, with the rotor stationary, the main flux might typically be only half what it was when the motor was at full speed. This is because at high slips, the leakage fluxes assume a much greater importance than under normal low-slip conditions. The simple arguments we have advanced to predict torque would therefore need to be modified to take account of the reduction of main flux if we wanted to use them quantitatively at high slips. There is no need for us to do this explicitly, but it will be reflected in any subsequent curves portraying typical torque-speed curves for real motors. Such curves are of course used when selecting a motor, since they provide the easiest means of checking whether the starting and run-up torque is adequate for the job in hand.

STATOR CURRENT-SPEED CHARACTERISTICS

In the previous section, we argued that as the slip increased, and the rotor did more mechanical work, the stator current increased. Since the extra current is primarily associated with the supply of real power (as distinct from the original magnetising current which was seen to be reactive), this work component of current is more or less in phase with the supply voltage, as shown in the phasor diagram, Figure 5.19.

The resultant stator current is the sum of the magnetizing current, which is present all the time, and the load component, which increases with the slip. We can see that as the load increases, the stator current also increases, and moves more nearly into phase with the voltage. But because the magnetizing current is appreciable, the difference in magnitude between no-load and full-load currents may not be

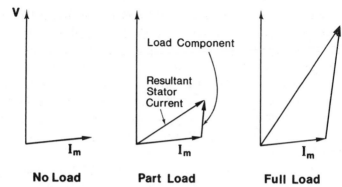

No Load   Part Load   Full Load

Figure 5.19   *Phasor diagrams showing stator current at no-load, part-load and full-load. The resultant current in each case is the sum of the no-load (magnetizing) current and the load component*

all that great. This is in sharp contrast to the d.c. motor, where the no-load current is very small in comparison with the full-load current.

Actually we cannot push the simple ideas behind Figure 5.19 too far. Strictly, they are an approximation, and whilst they are fairly close to the truth for the normal operating region, they break down at higher slips, where the rotor and stator leakage reactances become significant.

A typical current locus over the whole range of slips for a

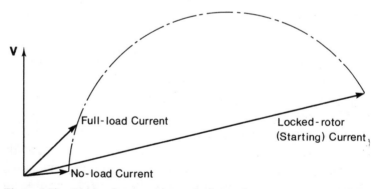

Figure 5.20   *Phasor diagram showing the locus of stator current over the full range of speeds from no-load (full speed) down to the locked-rotor (starting) condition*

cage motor is shown in Figure 5.20. We note that the power factor becomes worse again at high slips, and also that the current at standstill (i.e. the 'starting' current) is perhaps five times the rated value.

Very high starting currents are one of the worst features of the cage induction motor. They not only cause unwelcome volt-drops in the supply system, but also call for heavier switchgear than would be needed to cope with full-load conditions.

Unfortunately, for reasons discussed earlier, the high starting currents are not accompanied by high starting torques, as we can see from Figure 5.21, which shows current and torque as functions of slip.

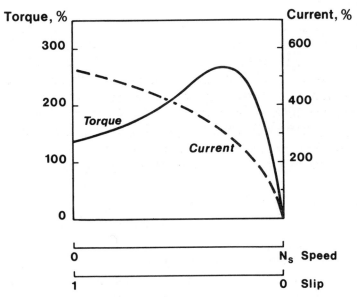

Figure 5.21 *Typical torque-speed and current-speed curves for a cage induction motor. The torque and current axes are scaled so that 100 per cent represents the continuously-rated (full-load) value*

In fact, the torque per ampere of current drawn from the mains is typically very low at start up, and only reaches a respectable value in the normal operating region, i.e. when the slip is small.

6

OPERATING CHARACTERISTICS OF INDUCTION MOTORS

This chapter is concerned with how the induction motor behaves when connected to a constant frequency supply. This remains by far the most widely used and important mode of operation, the motor running directly connected to a constant-voltage mains supply. Three-phase motors are the most important, so they are dealt with first.

METHODS OF STARTING CAGE MOTORS

We have seen that the starting current in a cage motor supplied at rated voltage is perhaps five or six times rated current, at a low power-factor. Where the impedance of the supply system is low, the volt-drop caused by even such a large current will be negligible, and other consumers on the same supply will be unaffected. No special arrangements are needed for starting the motor on such a 'stiff' supply, and the three motor leads are simply switched directly onto the mains. Starting in this way is known as 'direct-on-line' (DOL) or 'direct-to-line' (DTL). The switching will usually be done by means of a relay or contactor, incorporating fuses and other overload protection devices, and either operated manually by local or remote pushbuttons, or interfaced to permit operation from a programmable controller or computer.

In contrast, if the supply impedance is high, an appreciable volt-drop will occur every time the motor is started, causing lights to dim and interfering with other apparatus on the same supply. With this 'weak' supply, some form of starter is called for to limit the current at starting and during the run-up phase, thereby reducing the severity of the shock applied to the system. As the motor picks up speed, the current falls, so the starter is removed as the motor approaches full speed. Naturally enough the price to be paid for the reduction in current is a lower starting torque, and a longer run-up time.

Whether or not a starter is required depends on the size of the motor in relation to the capacity or 'fault-level' of the supply, the prevailing regulations imposed by the supply authority, and the nature of the load.

The references above to 'low' and 'high' supply impedances must therefore be interpreted in relation to the impedance of the motor when it is stationary. A large (and therefore low impedance) motor could well be started quite happily direct-on-line in a major industrial plant, where the supply is 'stiff' i.e. the supply impedance is very much less than the motor impedance. But the same motor would need a starter when used in a rural setting remote from the main power system, and fed by a relatively high impedance or 'weak' supply. Needless to say, the stricter the rules governing permissible volt-drop, the more likely it is that a starter will be needed.

Motors which start without significant load torque or inertia can accelerate very quickly, so the high starting current is only drawn for a short period. A 10 kW motor would be up to speed in a second or so, and the volt-drop may therefore be judged as acceptable. Clutches are sometimes fitted to permit 'off-load' starting, the load being applied after the motor has reached full speed. Conversely, if the load torque and/or inertia are high, the run-up may take many seconds, in which case a starter may prove essential. No strict rules can be laid down, but obviously the bigger the motor, the more likely it is to require a starter.

Star/delta (Wye/Mesh) starter

This is the simplest and most widely used method of starting. It provides for the windings of the motor to be connected in star (wye) to begin with, thereby reducing the voltage applied to each phase to 58 per cent ($1/\sqrt{3}$) of its direct-on-line value. Then, when the motor speed approaches its running value, the windings are switched to delta (mesh) connection. The main advantage of the method is its simplicity, while its main drawback is that the sudden transition from star to delta gives rise to a second shock – albeit of lesser severity – to the supply system and to the load. For star/delta switching to be possible both ends of each phase of the motor windings must be brought out to the terminal box. This requirement is met in the majority of motors, except small ones which are usually permanently connected in delta.

With a star/delta starter the current drawn from the supply is approximately one third of that drawn in a direct-on-line start, which is very welcome, but at the same time the starting torque is also reduced to one third of its direct-on-line value. Naturally we need to ensure that the reduced torque will be sufficient to accelerate the load, and bring it up to a speed at which it can be switched to delta without an excessive jump in the current.

Various methods are used to detect when to switch from star to delta. In manual starters, the changeover is determined by the operator watching the ammeter until the current has dropped to a low level, or listening to the sound of the motor until the speed becomes steady. Automatic versions are similar in that they detect either falling current or speed rising to a threshold level, or else they operate after a pre-set time.

Autotransformer starter

A three-phase autotransformer is usually used where it is necessary to reduce the starting current by less than that obtained with star-delta starting. Each phase of an autotransformer consists of a single winding on a laminated core. The mains supply is connected across the ends of the coils, and one or more tapping points (or a sliding contact) provide a reduced voltage output, as shown in Figure 6.1.

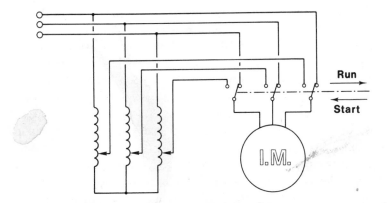

Figure 6.1   *Autotransformer starter for cage induction motor*

The motor is first connected to the reduced voltage output, and when the current has fallen to the running value, the motor leads are switched over to the full voltage.

If the reduced voltage is chosen so that a fraction a of the line voltage is used to start the motor, the starting torque is reduced to approximately a^2 times its direct-on-line value, and the current drawn from the mains is also reduced to a^2 times its direct value. As with the star/delta starter, the torque per ampere is the same as for a direct start.

The switchover from the starting tap to the full voltage inevitably results in mechanical and electrical shocks to the motor. In large motors the transient overvoltages caused by switching can be enough to damage the insulation, and where this is likely to pose a problem a modified procedure known as the Korndorfer method is used. A smoother changeover is achieved by leaving part of the winding of the autotransformer in series with the motor winding all the time.

Resistance or reactance starter

By inserting three resistors or inductors of appropriate value in series with the motor, the starting current can be reduced by any desired extent, but only at the expense of a disproportionate reduction in starting torque.

For example if the current is reduced to half its direct-on-line value, the motor voltage will be halved, so the torque (which is proportional to the square of the voltage – see later) will be reduced to only 25 per cent of its direct-on-line value. This approach is thus less attractive in terms of torque per ampere of supply current than the star/delta method. One attractive feature, however, is that as the motor speed increases and its effective impedance rises, the volt-drop across the extra impedance reduces, so the motor voltage rises progressively with the speed, thereby giving more torque. When the motor is up to speed, the added impedance is shorted-out by means of a contactor. Variable-resistance starters (manually or motor operated) are sometimes used with small motors where a smooth jerk-free start is required, for example in film or textile lines.

Solid-state soft starting

By the 1990s this method looks poised to overtake the star/delta method in terms of numbers of new installations. It provides a smooth build-up of current and torque, the maximum current and acceleration time are easily adjusted, and it is particularly valuable where the load must not be subjected to sudden jerks. The only real drawback over conventional starters is that the mains currents during run-up are not sinusoidal, which can lead to interference with other equipment on the same supply.

The most widely-used arrangement comprises three pairs of back-to-back thyristors connected in series with the three supply lines, as shown in Figure 6.2(a).

Each thyristor is fired once per half-cycle, the firing being synchronized with the mains and the firing angle being vari-

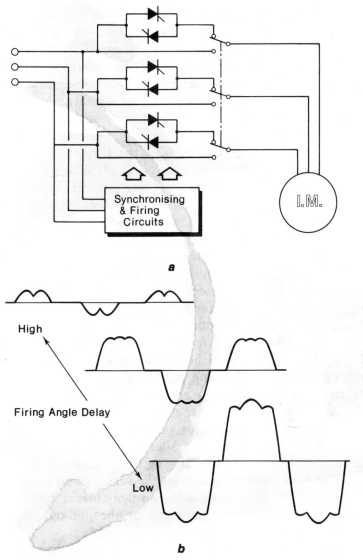

Figure 6.2 *(a) Thyristor soft-starter, (b) typical motor current waveforms*

able so that each pair conducts for a varying proportion of a cycle. Typical current waveforms are shown in Figure 6.2(b): they are clearly not sinusoidal but the motor will tolerate them quite happily.

A wide variety of control philosophies can be found, with the degree of complexity and sophistication being reflected in the price. The cheapest open-loop systems simply alter the firing angle linearly with time, so that the voltage applied to the motor increases as it accelerates. The 'ramp-time' can be set by trial and error to give an acceptable start, i.e. one in which the maximum allowable current from the supply is not exceeded at any stage. This approach is reasonably satisfactory when the load remains the same, but requires resetting each time the load changes. Loads with high static friction are a problem because nothing happens for the first part of the ramp, during which time the motor torque is insufficient to move the load. When the load finally moves, its acceleration is often too rapid. The more advanced open-loop versions allow the level of current at the start of the ramp to be chosen, and this is helpful with 'sticky' loads.

More sophisticated systems – usually with on-board digital controllers – provide for tighter control over the acceleration profile by incorporating closed-loop current feedback. After an initial ramping up to the start level (over the first few cycles), the current is held constant at the desired level throughout the accelerating period, the firing angle of the thyristors being continually adjusted to compensate for the changing effective impedance of the motor. By keeping the current at the maximum value which the supply can tolerate the run-up time is minimized. Alternatively, if a slow run-up is desirable, a lower accelerating current can be selected.

As with the open-loop systems the velocity-time profile is not necessarily ideal, since with constant current the motor torque exhibits a very sharp rise as the pull-out slip is reached, resulting in a sudden surge in speed.

Prospective users need to be wary of some of the promotional literature surrounding proprietary soft-start systems. In common with most relatively new products, the virtues tend to be exaggerated while the shortcomings are played down. Claims are sometimes made that massive reductions in starting current can be achieved without corresponding reductions in starting torque. This is nonsense. The current can certainly be limited, but as far as torque per line amp is concerned soft-start systems are no better than series reactor systems, and not as good as the autotransformer and star/delta methods.

RUN-UP AND STABLE OPERATING REGIONS

In addition to having sufficient torque to start the load it is obviously necessary for the motor to bring the load up to full speed. To predict how the speed will rise after switching-on we need the torque-speed curves of the motor and the load, and the total inertia.

By way of example, we can look at the case of a motor with two different loads (Figure 6.3). Load (A) is typical of a simple hoist, which applies constant torque to the motor at all speeds, while load (B) might represent a fan. For the sake of simplicity, we will assume that the load inertias (as seen at the motor shaft) are the same.

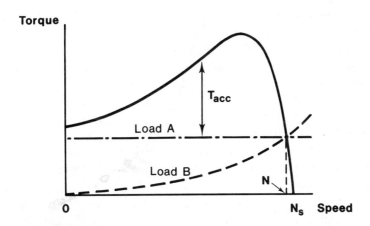

Figure 6.3   *Typical torque-speed curve showing two different loads which have the same steady running speed*

The speed-time curves for run-up are shown in Figure 6.4. Note that the gradient of the speed-time curve (i.e. the acceleration) is obtained by dividing the accelerating torque

T_{acc} (which is the difference between the torque developed by the motor and the torque required to run the load at that speed) by the total inertia.

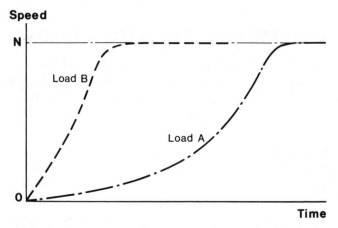

Figure 6.4   *Speed-time curves during run-up, for motor and loads shown in Figure 6.3*

In this example, both loads ultimately reach the same steady speed (i.e. the speed at which motor torque equals load torque), but B reaches full speed much more quickly because the accelerating torque is higher during most of the run-up. Load A picks up speed slowly at first, but then accelerates hard (with a characteristic 'whoosh') as it passes through the peak torque speed and approaches equilibrium conditions.

It should be clear that the higher the total inertia, the slower the acceleration, and vice-versa. The total inertia means the inertia as seen at the motor shaft, so if gearboxes or belts are employed the inertia must be 'referred' as discussed in Chapter 10.

High inertia loads – overheating

Apart from accelerating slowly, heavy inertia loads pose a particular problem of rotor heating which can easily be overlooked by the unwary user. Every time an induction motor is started from rest and brought up to speed, the total energy dissipated as heat in the motor windings is equal to the stored kinetic energy of the motor plus load. Hence with high inertia loads, very large amounts of energy are released as heat in the windings during run-up, even if the load torque is negligible when the motor is up to speed. With totally-enclosed motors the heat ultimately has to find its way to the finned outer casing of the motor, which is cooled by air from the shaft-mounted external fan. Cooling of the rotor is there-fore usually much worse than the stator, and the rotor is thus most likely to overheat during high inertia run-ups.

No hard and fast rules can be laid down, but manufac-turers usually work to standards which specify how many starts per hour can be tolerated. Actually, this information is useless unless coupled with reference to the total inertia, since doubling the inertia makes the problem twice as bad. However, it is usually assumed that the total inertia is not likely to be more than twice the motor inertia, and this is certainly the case for most loads, but if in doubt, the user should consult the manufacturer who may recommend a larger motor than might seem necessary simply to supply the full-load power requirements.

Steady-state rotor losses and efficiency

The discussion above is a special case which highlights one of the less attractive features of induction machines. This is that it is never possible for all the power crossing the air-gap to be converted to mechanical output, because some is always lost as heat in the rotor resistance. In fact, it turns out that at slip s the total power (P_r) crossing the air-gap always divides so that sP_r is lost as heat, while $(1 - s)P_r$ is converted to useful mechanical output.

When the motor is operating in the steady-state the energy-conversion efficiency of the rotor is thus given by

$$\eta_r = \frac{\text{output}}{\text{input}} = (1 - s).$$

This result is very important, and shows us immediately why operating at small values of slip is desirable. With a slip of 5 per cent (or 0.05) for example, 95 per cent of the air-gap power is put to good use. But if the motor was run at half the synchronous speed (s = 0.5), 50 per cent of the air-gap power would be wasted as heat in the rotor.

We can also see that the overall efficiency of the motor must always be significantly less than $(1 - s)$, because in addition to the rotor copper losses there are stator copper losses, iron losses and windage and friction losses. This fact is sometimes forgotten, leading to conflicting claims such as 'full-load slip = 5 per cent, overall efficiency = 96 per cent', which is clearly impossible.

Steady-state stability – pull-out torque and stalling

We can check stability by asking what happens if the load torque suddenly changes for some reason. The load shown in Figure 6.5 is stable at speed X, for example: if the load torque increased from T_a to T_b, the load torque would be greater than the motor torque, so the motor would decelerate. As the speed dropped, the motor torque would rise,

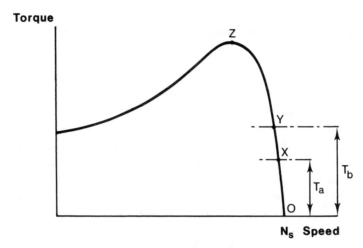

Figure 6.5 *Torque-speed curve illustrating stable operating region (OXYZ)*

until a new equilibrium was reached, at the slightly lower speed (Y). The converse would happen if the load torque reduced, leading to a higher stable running speed.

But what happens if the load torque is increased more and more? We can see that as the load torque increases, beginning at point X, we eventually reach point Z, at which the motor develops its maximum torque. Quite apart from the fact that the motor is now well into its overload region, and will be in danger of overheating, it has also reached the limit of stable operation. If the load torque is further increased, the speed falls (because the load torque is more than the motor torque), and as it does so the shortfall between motor torque and load torque becomes greater and greater. The speed therefore falls faster and faster, and the motor is said to be 'stalling'. With loads such as machine tools (a drilling machine, for example), as soon as the maximum or 'pull-out' torque is exceeded, the motor rapidly comes to a halt, making an angry humming sound. With a hoist, however, the excess load would cause the rotor to be accelerated in the reverse direction, unless it was prevented from doing so by a mechanical brake.

TORQUE-SPEED CURVES – INFLUENCE OF ROTOR PARAMETERS

We saw earlier that the rotor resistance and reactance influenced the shape of the torque-speed curve. Both of these parameters can be varied by the designer, and we will explore the pros and cons of the various alternatives. To limit the mathematics the discussion will be mainly qualitative, but it is worth mentioning that the whole matter can be dealt with rigorously using the equivalent circuit approach.

We will deal with the cage rotor first because it is the most important, but the wound rotor allows a wider variation of resistance to be obtained, so it is discussed later.

Cage rotor

For small values of slip, i.e. in the normal running region, the lower we make the rotor resistance the steeper the slope of the torque-speed curve becomes, and the higher the rotor efficiency (which is equal to 1 − s), as shown in Figure 6.6.

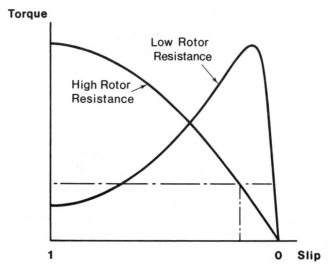

Figure 6.6   *Influence of rotor resistance on torque-speed curve of cage motor*

Both of these effects are desirable in most applications, because the speed will drop less as the load increases and the power wasted in the rotor will be minimized. There is of course a limit to how low we can make the resistance: copper allows us to achieve a lower resistance than aluminium, but we can't do any better than fill the slots with solid copper bars.

As we might expect there are drawbacks with a low resistance rotor. The starting torque is reduced (see Figure 6.6), and worse still the starting current is increased. The lower starting torque may prove insufficient to accelerate the load, while increased starting current may lead to unacceptable volt-drops in the supply.

Altering the rotor resistance has little or no effect on the value of the peak (pull-out) torque, but the slip at which the peak torque occurs is directly proportional to the rotor resistance. By opting for a high enough resistance (by making the cage from bronze, brass or other high-resistivity material) we can if we wish arrange for the peak torque to occur at starting, as shown in Figure 6.6. The snag in doing this is that the full-load efficiency is inevitably low because the full-load slip will be high.

There are some applications for which high-resistance motors are well suited, an example being for metal punching presses, where the motor accelerates a flywheel which is used to store energy. In order to release a significant amount of energy, the flywheel slows down appreciably during impact, and the motor then has to accelerate it back up to full speed. The motor needs a high torque over a comparatively wide speed range, and does most of its work during acceleration. Once up to speed the motor is effectively running light, so its low efficiency is of little consequence. High-resistance motors are also beginning to be widely used for speed control of fan-type loads, and this is taken up again in the section on Speed Control.

To sum up, a high rotor resistance is desirable when starting and at low speeds, while a low resistance is preferred under normal running conditions. To get the best of both worlds, we need to be able to alter the resistance from a high value at starting to a lower value at full speed. Obviously we can't change the actual resistance of the cage once it has been manufactured, but it is possible to achieve the desired effect with either a 'double cage' or a 'deep bar' rotor. Manufacturers normally offer a range of designs which reflect these trade-offs, and the user then selects the one which best meets his particular requirements.

Double cage rotors

Double cage rotors have an outer cage made of relatively high resistivity material such as bronze, and an inner cage of low resistivity, usually copper, as shown in Figure 6.7.

 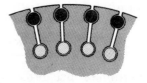

Figure 6.7 *Alternative arrangements of double-cage rotors. The outer cage has a high resistance (e.g. bronze) while the inner cage has a low resistance (e.g. copper)*

The inner cage is sunk deep into the rotor, so that it is almost completely surrounded by iron. This causes the inner bars to have a much higher leakage inductance than if they were near the rotor surface, so that under starting conditions their inductive reactance is very high and little current flows in them. In contrast, the bars of the outer cage are placed so that their leakage fluxes face a much higher reluctance path, leading to a low leakage inductance. Hence under starting conditions, rotor current is concentrated in the outer cage, which, because of its high resistance, produces a high starting torque.

At the normal running speed the roles are reversed. The rotor frequency is low, so both cages have low reactance and most of the current therefore flows in the low-resistance inner cage. The torque-speed curve is therefore steep, and the efficiency is high.

Considerable variation in detailed design is possible in order to shape the torque-speed curve to particular requirements. In comparison with a single cage rotor, the double cage gives much higher starting torque, substantially less starting current, and marginally worse running performance.

Deep bar rotors

The deep bar rotor has a single cage, usually of copper, formed in slots which are deeper and narrower than in a conventional single-cage design. Construction is simpler and therefore cheaper than in a double-cage rotor as shown in Figure 6.8.

Figure 6.8 *Typical deep-bar rotor construction*

The deep bar approach ingeniously exploits the fact that the effective resistance of a conductor is higher under a.c. conditions than under d.c. conditions. With a typical copper bar of the size used in an induction motor rotor, the difference in effective resistance between d.c. and say 50 or 60 Hz (the so-called 'skin-effect') would be negligible if the conductor were entirely surrounded by air. But when it is almost completely surrounded by iron, as in the rotor slots, its effective resistance at mains frequency may be two or three times its d.c. value.

Under starting conditions, when the rotor frequency is equal to the supply frequency, the skin effect is very pronounced, and the rotor current is concentrated towards the top of the slots. The effective resistance is therefore increased, resulting in a high starting torque from a low starting current. When the speed rises and the rotor frequency falls, the effective resistance reduces towards its d.c. value, and the current distributes itself more uniformly across the cross-section of the bars. The normal running performance thus approaches that of a low-resistance single-cage rotor, giving a high efficiency and stiff torque-speed curve. The pull-out torque is however somewhat lower than for an equivalent single-cage motor because of the rather higher leakage reactance.

Most small and medium motors are designed to exploit the deep bar effect to some extent, reflecting the view that for most applications the slightly inferior running performance is more than outweighed by the much better starting behaviour. A typical torque-speed curve for a general-purpose medium-size (55 kW) motor is shown in Figure 6.9.

Such motors are unlikely to be described by the maker specifically as deep-bar, but they nevertheless incorporate a measure of the skin effect and consequently achieve the good torque-speed characteristic shown in Figure 6.9.

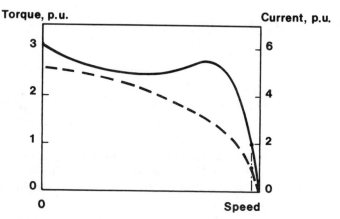

Figure 6.9   *Complete torque-speed and current-speed curves for a general-purpose industrial cage motor*

The current-speed relationship is also shown in Figure 6.9, and both torque and current scales are expressed in per-unit (p.u.). This notation is widely used as a shorthand, with 1 p.u. (or 100 per cent) representing rated value. For example a torque of 1.5 p.u. simply means one and a half times rated value, while a current of 400 per cent means a current of four times rated value.

Starting and run-up of slipring motors

By adding external resistance in series with the rotor windings the starting current can be kept low but at the same time the starting torque is high. This is the major advantage of the wound-rotor or slipring motor, and makes it well suited for loads with heavy starting duties such as stone-crushers, cranes and conveyor drives.

The influence of rotor resistance is shown by the set of torque-speed curves in Figure 6.10.

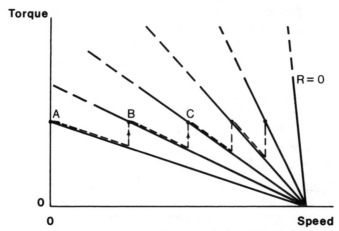

Figure 6.10 *Torque-speed curves for a wound-rotor (slipring) motor showing how the rotor circuit resistance can be varied in steps to provide an approximately constant torque during acceleration*

Plate 6.1 *Drip proof wound-rotor induction motor, 22 kW at 960 rev/min. External air is drawn in at one end to cool the stator and rotor windings before being expelled at the other end. A finned case is not required, and the sliprings are located beneath the cover at the non-drive end (Photograph by courtesy of Brook Crompton Parkinson Motors)*

A high rotor resistance is used when the motor is first switched on, and depending on the value chosen any torque up to the pull-out value (perhaps twice full-load) can be obtained. Typically the resistance will be selected to give full-load torque at starting, together with rated current from the mains. The starting torque is then as indicated by point A in Figure 6.10.

As the speed rises, the torque would fall more or less linearly if the resistance remained constant, so in order to keep close to full torque the resistance is gradually reduced, either in steps, in which case the trajectory ABC etc. is followed (Figure 6.10), or continuously so that maximum torque is obtained throughout. Ultimately the external resistance is made zero by shorting-out the sliprings, and thereafter the motor behaves like a low-resistance cage motor, with a high running efficiency.

As mentioned earlier, the total energy dissipated in the rotor circuit during run-up is equal to the final stored kinetic energy of the motor and load. In a cage motor this energy ends up in the rotor, and can cause overheating. In the slipring motor, however, most of the energy goes into the external resistance. This is a good thing from the motor point of view, but means that the external resistance has to absorb the thermal energy without overheating.

Fan-cooled grid resistors are often used, with tappings at various resistance values. These are progressively shorted-out during run-up, either by a manual or motor-driven drum-type controller, or with a series of timed contactors. Alternatively, where stepless variation of resistance is required, a liquid resistance controller is often employed. It consists of a tank of electrolyte (typically caustic soda) into which three electrodes can be raised or lowered. The resistance between the electrodes depends on how far they are immersed in the liquid. The electrolyte acts as an excellent short-term reservoir for the heat released, and by arranging for convection to take place via a cooling radiator, the equipment can also be used continuously for speed control (see later).

Attempts have been made to vary the effective rotor circuit resistance by means of a fixed external resistance and a set of series connected thyristors, but this approach has not gained wide acceptance.

INFLUENCE OF SUPPLY VOLTAGE ON TORQUE-SPEED CURVE

We established earlier that at any given slip, the air-gap flux is proportional to the applied voltage, and the induced current in the rotor is proportional to the flux. The torque – which depends on the product of the flux and the rotor current – therefore depends on the square of the applied voltage. This means that a comparatively modest fall in the voltage will result in a much larger reduction in torque capability, with adverse effects which may not be apparent to the unwary until too late.

To illustrate the problem, consider the torque-speed curves for a cage motor shown in Figure 6.11. The curves (which have been expanded to focus attention on the low-slip region) are drawn for full voltage (100 per cent), and for a modestly reduced voltage of 90 per cent. With full voltage and full-load torque the motor will run at point X, with a slip of say 5 per cent. Since this is the normal full-load

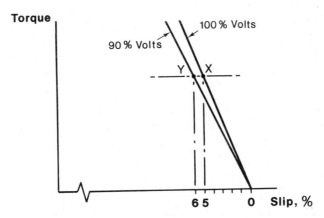

Figure 6.11 *Influence of stator supply voltage on torque-speed curves*

condition, the rotor and stator currents will be at their rated values.

Now suppose that the voltage falls to 90 per cent. The load torque is assumed to be constant so the new operating point will be at Y. Since the air-gap flux is now only 0.9 of its rated value, the rotor current will have to be about 1.1 times rated value to develop the same torque, so the rotor e.m.f. is required to increase by 10 per cent. But the flux has fallen by 10 per cent, so an increase in slip of 20 per cent is called for. The new slip is therefore 6 per cent.

The drop in speed from 95 per cent of synchronous to 94 per cent may well not be noticed, and the motor will apparently continue to operate quite happily. But the rotor current is now 10 per cent above its rated value, so the rotor heating will be 21 per cent more than is allowable for continuous running. The stator current will also be above rated value, so if the motor is allowed to run continuously, it will overheat. This is one reason why all large motors are fitted with protection which is triggered by over-temperature. Many small and medium motors do not have such protection, so it is important to guard against the possibility of undervoltage operation.

GENERATING AND BRAKING

Having explored the torque-speed curve for the normal motoring region, where the speed lies between zero and just below synchronous, we must ask what happens if the speed is above the synchronous speed, or is negative.

A typical torque-speed curve for a cage motor covering the full range of speeds which are likely to be encountered in practice is shown in Figure 6.12.

We can see from Figure 6.12 that the decisive factor as far as the direction of the torque is concerned is the slip, rather than the speed. When the slip is positive the torque is positive, and vice-versa. The torque therefore always acts so as to urge the rotor to run at zero slip, i.e. at the synchronous speed. If the rotor is tempted to run faster than the field it

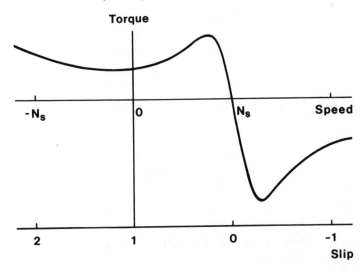

Figure 6.12   *Torque-speed curve over motoring region (slip between 0 and 1), braking region (slip greater than 1) and generating region (slip negative)*

will be slowed down, whilst if it is running below synchronous speed it will be accelerated forwards. In particular, we note that for slips greater than 1, i.e. when the rotor is turning in the opposite direction to the field, the torque will remain positive, so that if the rotor is unrestrained it will first slow down and then change direction and accelerate in the direction of the field.

Generating region – overhauling loads

For negative slips, i.e. when the rotor is turning in the same direction, but at a higher speed than the travelling field, the 'motor' torque is in fact negative. In other words the machine develops a torque which opposes the rotation, which can therefore only be maintained by applying a driving torque to the shaft. In this region the machine acts as an induction generator, converting mechanical power from the shaft into electrical power into the supply system.

Before we look at why this is important to the motor user,

we should be clear that the machine can only generate when it is connected to the supply. If we disconnect an induction motor from the mains and try to make it generate simply by turning the rotor we will not get any output because there is nothing to set up the working flux. The flux or excitation is not present until the motor is supplied with magnetizing current from the supply.

There are comparatively few situations in which mains-fed motors find themselves in the generating region, though as we will see later it is quite common in inverter-fed applications. We will however look at one example of a mains fed motor in the so-called 'regenerative' mode to underline the value of the motor's inherent ability to switch from motoring to generating automatically, without the need for any external intervention.

Consider a cage motor driving a simple hoist through a reduction gearbox, and suppose that the hook (unloaded) is to be lowered. Because of the static friction in the system, the hook will not descend on its own, even after the brake is lifted, so on pressing the 'down' button the brake is lifted and power is applied to the motor so that it rotates in the lowering direction. The motor quickly reaches full speed and the hook descends. As more and more rope winds off the drum, a point is reached where the lowering torque exerted by the hook and rope is greater than the running friction, and a restraining torque is then needed to prevent a runaway. The necessary stabilizing torque is automatically provided by the motor acting as a generator as soon as the synchronous speed is exceeded, as shown in Figure 6.12. The speed will therefore be held at just above the synchronous speed, provided of course that the peak generating torque (see Figure 6.12) is not exceeded.

Plug reversal and plug braking

Because the rotor always tries to catch up with the rotating field, it can be reversed rapidly simply by interchanging any two of the supply leads. The changeover is usually obtained by having two separate three-pole contactors, one for forward and one for reverse. This procedure is known as plug reversal or plugging, and is illustrated in Figure 6.13.

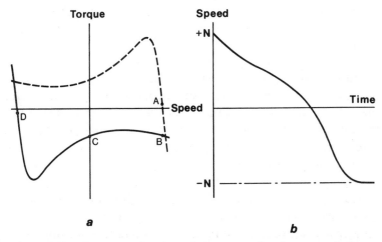

Figure 6.13   *Torque-speed and speed-time curves for plug reversal of cage motor*

The motor is initially assumed to be running light (and therefore with a very small positive slip) as indicated by point A on the dotted torque-speed curve in Figure 6.13(a). Two of the supply leads are then reversed, thereby reversing the direction of the field, and bringing the mirror-image torque-speed curve shown by the solid line into play. The slip of the motor immediately after reversal is approximately 2, as shown by point B on the solid curve. The torque is thus negative, and the motor decelerates, the speed passing through zero at point C and then rising in the reverse direction before settling at point D, just below the synchronous speed.

The speed-time curve is shown in Figure 6.13(b). We can see that the deceleration (i.e. the gradient of the speed-time graph) gets progressively steeper until the motor passes through the peak torque (pull-out) point, but thereafter the final speed is approached gradually, as the torque tapers down to point D.

Very rapid reversal is possible using plugging; for example a 1 kW motor will typically reverse from full speed in under one second. But large cage motors can only be plugged if the supply can withstand the very high currents involved, which are even larger than when starting from rest. Frequent plugging will also cause serious overheating, because each reversal involves the 'dumping' of four times the stored kinetic energy as heat in the windings.

Plugging can be used to stop the rotor quickly, but obviously it is then necessary to disconnect the supply when the rotor comes to rest, otherwise it will run up to speed in reverse. A shaft-mounted reverse-rotation detector is therefore used to trip out the reverse contactor when the speed reaches the point of reversal.

We should note that whereas in the regenerative mode (discussed in the previous section) the slip was negative, allowing mechanical energy from the load to be converted to electrical energy and fed back to the mains, plugging is a wholly dissipative process in which all the kinetic energy ends up as heat in the motor.

Injection braking

This is the most widely-used method of electrical braking. When the 'stop' button is pressed, the three-phase supply is interrrupted, and a d.c. current is fed into the stator via two of its terminals. The d.c. supply is usually obtained from a rectifier fed via a low-voltage high-current transformer.

We saw earlier that the speed of rotation of the air-gap field is directly proportional to the supply frequency, so it should be clear that since d.c. is effectively zero frequency, the air-gap field will be stationary. We also saw that the rotor always tries to run at the same speed as the field. So if the field is stationary, and the rotor is not, a braking torque will be exerted. A typical torque-speed curve for braking a cage motor is shown in Figure 6.14, from which we see that the braking (negative) torque falls to zero as the rotor comes to rest.

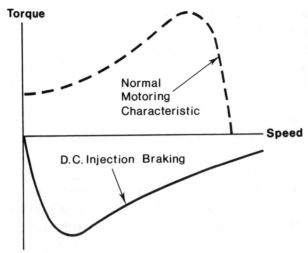

Figure 6.14   *Torque-speed curve for d.c. injection braking of cage motor*

This is in line with what we would expect, since there will only be induced currents in the rotor (and hence torque) when the rotor is 'cutting' the flux. As with plugging, injection (or dynamic) braking is a dissipative process, all the kinetic energy being turned into heat inside the motor.

SPEED CONTROL

We have seen that to operate efficiently an induction motor must run with a small slip. It follows that any efficient method of speed control must be based on varying the synchronous speed of the field, rather than the slip. The two factors which determine the speed of the field are the supply frequency and the pole-number (see equation 5.1).

The pole-number has to be an even integer, so where continuously adjustable speed control over a wide range is called for, the best approach is to provide a variable-frequency supply. This method is very important, and is dealt with separately in Chapter 7. But in this chapter we are concerned with constant frequency mains operation, so we have a choice between pole-changing, which can provide discrete speeds only, or slip-control which can provide continuous speed control, but is inherently inefficient.

Pole-changing motors

For some applications continuous speed control may be an unnecessary luxury, and it may be sufficient to be able to run at two discrete speeds. Among many instances where this can be acceptable and economic are pumps, fans and some machine tool drives.

We established in Chapter 5 that the pole-number of the field was determined by the layout of the stator coils, and that once the winding has been designed, and the frequency specified, the synchronous speed of the field is fixed. If we wanted to make a motor which could run at either of two different speeds, we could construct it with two separate stator windings (say 4-pole and 6-pole), and energize the appropriate one. There is no need to change the cage rotor since the pattern of induced currents can readily adapt to suit the stator pole-number. Early two-speed motors did have two distinct stator windings, but were bulky and inefficient.

It was soon realized that if half of the phase-belts within each phase-winding could be reversed in polarity, the effective pole-number could be halved. For example a 4-pole MMF pattern (N-S-N-S) would become (N-N-S-S), i.e. effectively a 2-pole pattern with one large N and one large S pole. By bringing out six leads instead of three, and providing switching contactors to effect the reversal, two discrete speeds in the ratio 2:1 are therefore possible from a single winding. The performance at the high (e.g. 2-pole) speed is relatively poor, which is not surprising in view of the fact that the winding was originally optimized for 4-pole operation.

It was not until the advent of the more sophisticated Pole Amplitude Modulation (PAM) method in the 1960s that two-speed single-winding high-performance motors with more or less any ratio of speeds became available from manufacturers. This subtle technique allows close ratios such as 4/6, 6/8, 8/10 or wide ratios such as 2/24 to be achieved. Close ratios are used in pumps and fans, while wide ratios are used for example in washing machines where a fast spin is called for.

The beauty of the PAM method is that it is not expensive. The stator winding has more leads brought out, and the coils are connected to form non-uniform phase-belts, but otherwise construction is the same as for a single-speed motor. Typically six leads will be needed, three of which are supplied for one speed, and three for the other, the switching being done by contactors. The method of connection (star or delta) and the number of parallel paths within the winding are arranged so that the air-gap flux at each speed matches the load requirement. For example if constant torque is needed at both speeds, the flux needs to be made the same, whereas if reduced torque is acceptable at the higher speed the flux can obviously be lower.

Voltage control of high-resistance cage motors

Where efficiency is not of paramount importance, the torque (and hence the running speed) of a cage motor can be controlled simply by altering the supply voltage. The torque at any slip is approximately proportional to the square of the voltage, so we can reduce the speed of the load by reducing the voltage. The method is not suitable for standard low-resistance cage motors, because their stable operating speed range is very restricted, as shown in Figure 6.15(a). But if

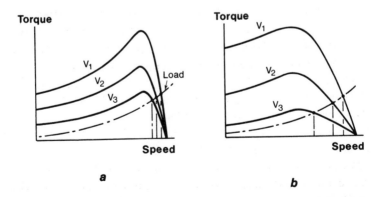

Figure 6.15   *Speed control of cage motor by stator voltage variation (a) low resistance rotor; (b) high resistance rotor*

special high-rotor-resistance motors are used, the slope of the torque-speed curve in the stable region is much less, and a wider range of steady-state operating speeds is available, as shown in Figure 6.15(b).

The most unattractive feature of this method is the low efficiency which is inherent in any form of slip-control. We recall that the rotor efficiency at slip s is $(1 - s)$, so if we run at say 70 per cent of synchronous speed (i.e. $s = 0.3$), 30 per cent of the power crossing the air-gap is wasted as heat in the rotor conductors. The approach is therefore only practicable where the load torque is low at low speeds, so that at high slips the heat in the rotor is tolerable. A fan-type characteristic is suitable, as shown in Figure 6.15(b), and many ventilating systems therefore use voltage-control.

Voltage control has only become worth considering since relatively cheap thyristor a.c. voltage regulators arrived on the scene during the 1970s. Previously the cost of auto-transformers or induction regulators to obtain the variable voltage supply was simply too high. The thyristor hardware required is essentially the same as discussed earlier for soft starting, and a single piece of kit can therefore serve for both starting and speed control. Where accurate speed control is needed, a tachogenerator must be fitted to the motor to provide a speed feedback signal, and this naturally increases the cost significantly.

Applications are now quite numerous, mainly in the range 0.5–10 kW, and most motor manufacturers offer high-resistance motors specifically for use with thyristor regulators. This relatively recent upsurge echoes the once widespread use of very small (and therefore high-resistance) induction motors as a.c. servo drives.

Speed control of wound-rotor motors

The fact that the rotor resistance can be varied easily allows us to control the slip from the rotor side, with the stator supply voltage and frequency constant. Although the method is inherently inefficient it is still used in many medium and large drives such as hoists, conveyors and crushers because of its simplicity and comparatively low cost.

A complete set of torque-speed characteristics is shown in Figure 6.16, from which it should be clear that by appropriate selection of the rotor circuit resistance, any torque up to typically 1.5 times full-load torque can be achieved at any speed.

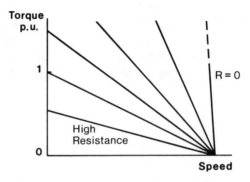

Figure 6.16 *Influence of external rotor resistance on torque-speed curve of wound-rotor motor*

POWER-FACTOR CONTROL AND ENERGY OPTIMIZATION

In addition to their use for soft-start and speed control, thyristor voltage regulators are often marketed as power-factor controllers and/or energy optimizers for cage motors. Some of the hype surrounding their introduction has thankfully evaporated, but users should remain sceptical of some of the more extravagent claims which can still be found.

The fact is that there are comparatively few situations where these considerations alone are sufficient to justify the expense of a voltage controller. Only when the motor operates for very long periods running light or at low load can sufficient savings be made to cover the outlay. There is certainly no point in providing energy-economy when the motor spends most of its time working at or near full-load.

Both power-factor control and energy optimization rely on the fact that the air-gap flux is proportional to the supply voltage, so that by varying the voltage, the flux can be set at the best level to cope with the prevailing load. We can see straightaway that nothing can be achieved at full load, since the motor needs full flux (and hence full voltage) to operate as intended. Some modest savings in losses can be achieved at reduced load, as we will see.

If we imagine the motor to be running with a low load torque and full voltage, the flux will be at its full value, and the magnetizing component of the stator current will be larger than the work component, so the input power-factor ($\cos\phi_a$) will be very low, as shown in Figure 6.17(a).

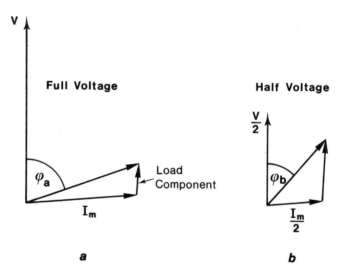

Figure 6.17 *Phasor diagram showing improvement of power-factor by reduction of stator voltage*

Now suppose that the voltage is reduced to say half (by phasing back the thyristors), thereby halving the air-gap flux and reducing the magnetizing current by at least a factor of two. With only half the flux, the rotor current must double to produce the same torque, so the work current reflected in the stator will also double. The input power-factor ($\cos\phi_b$) will therefore improve considerably (see Figure 6.17(b)). Of course the slip with 'half-flux' operation will be higher (by a factor of four), but with a low resistance cage it will still be small, and the drop in speed will therefore be slight.

The success (or otherwise) of the energy economy obtained depends on the balance between the iron losses and the copper losses in the motor. Reducing the voltage reduces the flux, and hence reduces the eddy current and hysteresis losses in the iron core. But as we have seen above, the rotor current has to increase to produce the same torque, so the rotor copper loss rises. The stator copper loss falls a little, because the power factor improves. In practice, with average general purpose motors, a nett saving in losses only occurs for light loads, say at or below 25 per cent of full load.

Slip energy recovery (wound-rotor motors)

Instead of wasting power in an external resistance, it can be returned to the mains supply. Direct connection is not possible because the rotor currents are at slip frequency. So the slip-frequency a.c. from the rotor is first rectified in a three-phase diode bridge and smoothed before being returned to the mains supply via a three-phase thyristor bridge converter. A transformer is usually required to match the output from the controlled bridge to the mains voltage.

Since the cost of the converters depends on the slip power they have to handle, this system (which is known as the static Kramer drive) is most often used where only a modest range of speeds (say from 80 per cent of synchronous and above) is required, such as in large pump and compressor drives. Speed control is obtained by varying the firing angle of the controlled converter, the torque-speed curves for each firing angle being fairly steep (i.e. approximating to constant speed), thereby making closed-loop speed control relatively simple.

SINGLE-PHASE INDUCTION MOTORS

Single-phase induction motors are simple, robust and reliable, and are used in enormous numbers especially in domestic and commercial applications where three-phase supplies are not available. Although outputs of up to a few kW are possible, the majority are below 0.5 kW, and are used in such applications as refrigeration compressors, washing machines and dryers, pumps and fans, small machine tools, tape and record decks, printing machines, etc.

Principle of operation

If one of the leads of a 3-phase motor is disconnected while it is running light, it will continue to run with a barely perceptible drop in speed, and a somewhat louder hum. With only two leads remaining there can only be one current, so the motor must be operating as a single-phase machine. If load is applied the slip increases more quickly than under three-phase operation, and the stall torque is much less, perhaps one-third. When the motor stalls and comes to rest it will not restart if the load is removed, but remains at rest drawing a heavy current and emitting an angry hum. It will burn out if not disconnected rapidly.

It is not surprising that a truly single-phase cage induction motor will not start from rest, because the single winding, fed with a.c., simply produces a pulsating flux in the air-gap, without any suggestion of rotation. It is however surprising to find that if the motor is given a push in either direction it will pick up speed, slowly at first but then with more vigour, until it settles with a small slip, ready to take up load. Once turning, a rotating field is evidently brought into play to continue propelling the rotor.

We can understand how this comes about by first picturing the pulsating MMF set up by the current in the stator winding as being the resultant of two identical travelling waves of MMF, one in the forward direction and the other in reverse. When the rotor is stationary, it reacts equally to both waves, and no torque is developed. When the rotor is turning, however, the induced rotor currents are such that their MMF opposes the reverse stator MMF to a greater extent than they oppose the forward stator MMF. The result is that the forward flux wave (which is what develops the forward torque) is bigger than the reverse flux wave (which exerts a drag). The difference widens as the speed increases, the forward flux becoming progressively bigger as the speed rises while the reverse flux simultaneously reduces. This 'positive feedback' effect explains why the speed builds slowly at first, but later zooms up to just below synchronous speed. At the normal running speed (i.e. small slip), the forward flux is many times larger than the reverse flux, and the drag torque is only a small percentage of the forward torque.

As far as normal running is concerned, a single winding is therefore sufficient. But all motors must be able to self-start, so some mechanism has to be provided to produce a rotating field even when the rotor is at rest. Several methods are employed, all of them using an additional winding.

The second winding usually has less copper than the main winding, and is located in the slots which are not occupied by the main winding, so that its MMF is displaced in space relative to that of the main winding. The current in the second winding is supplied from the same single-phase source as the main winding current, but is caused to have a phase-lag, by a variety of means which are discussed later. The combination of a space displacement between the two windings together with a time displacement between the currents produces a two-phase machine. If the two windings were identical, displaced by 90°, and fed with currents with 90° phase-shift, an ideal rotating field would be produced. In practice we can never achieve a 90° phase-shift between the currents, and it turns out to be more economical not to make the windings identical. Nevertheless, a decent rotating field is set up, and entirely satisfactory starting torque can be obtained. Reversal is simply a matter of reversing the polarity of one of the windings, and performance is identical in both directions.

The most widely used starting methods are described below. At one time it was common practice for the second or auxiliary winding to be energized only during start and run-up, and for it to be disconnected by means of a centrifugal switch mounted on the rotor, or sometimes by a time-switch. This practice gave rise to the term 'starting winding'. Nowadays it is more common to find both windings in use all the time.

Capacitor-run motors

A capacitor is used in series with the auxiliary winding (Figure 6.18(a)) to provide a phase-shift between the main and auxiliary winding currents. The capacitor (usually of a few μF, and with a voltage rating which may well be higher than the mains voltage) may be mounted piggyback fashion on the motor, or located elsewhere. Its value represents a compromise between the conflicting requirements of high starting torque and good running performance.

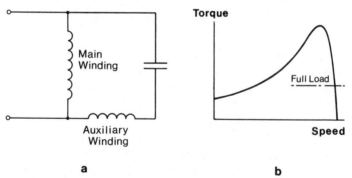

a   b

Figure 6.18   *Single-phase capacitor-run induction motor*

A typical torque-speed curve is shown in Figure 6.18 (b); the modest starting torque indicates that capacitor-run motors are generally best suited to fan-type loads. Where higher starting torque is needed, two capacitors can be used, one being switched out when the motor is up to speed.

As mentioned above, the practice of switching out the

Plate 6.2 *Fractional-horsepower cage induction motors. Those with piggyback capacitors are recognizable as single-phase, while those without are either three-phase or split-phase. Most mass-produced motors in these low powers are 'special' in some way, being tailored to meet customer requirements (Photograph by courtesy of GEC Electromotors Ltd)*

starting winding altogether is no longer favoured for new machines, but many old ones remain, and where a capacitor is used they are known as 'capacitor start' motors.

Split-phase motors

The main winding is of thick wire, with a low resistance and high reactance, while the auxiliary winding is made of fewer turns of thinner wire with a higher resistance and lower reactance (Figure 6.19(a)). The inherent difference in impedance is sufficient to give the phase-shift between the two currents without needing any external elements in series. Starting torque is good at typically 1.5 times full-load torque, as shown in Figure 6.19(b). As with the capacitor type, reversal is accomplished by changing the connections to one of the windings.

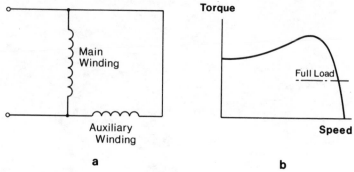

a   b

Figure 6.19   *Single-phase split-phase induction motor*

Shaded-pole motors

There are several variants of this extremely simple, robust and reliable cage motor, which predominates for low-power applications such as hair-dryers, oven fans, tape decks, office equipment, display drives etc. A common 2-pole version from the cheap end of the market is shown in Figure 6.20.

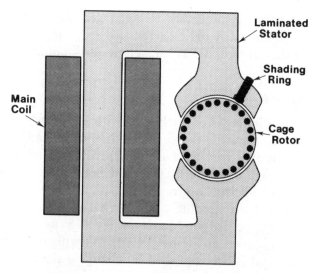

Figure 6.20   *Shaded-pole induction motor*

The rotor, typically between 1 and 4 cm diameter, has a die-cast aluminium cage, while the stator winding is a simple concentrated coil wound round the laminated core. The stator pole is slotted to receive the 'shading ring' which is a single short-circuited turn of thick copper or aluminium.

Most of the pulsating flux produced by the stator winding bypasses the shading ring and crosses the air-gap to the rotor. But some of the flux passes through the shading ring, and because it is alternating it induces an e.m.f. and current in the ring. The opposing MMF of the ring current diminishes and retards the phase of the flux through the ring, so that the flux through the ring reaches a peak after the main flux, thereby giving what amounts to a rotation of the flux across the face of the pole. This far from perfect travelling wave of flux produces the motor torque by interaction with the rotor cage. Efficiencies are low because of the rather poor

Plate 6.3   *Single-phase shaded-pole motor. These motors are produced in very large numbers for use in small domestic fans, office equipment etc (Photograph by courtesy of Brook Crompton Parkinson Motors)*

magnetic circuit and the losses caused by the induced currents in the shading ring, but this is generally acceptable when the aim is to minimize first cost. Series resistance can be used to obtain a crude speed control, but this is only suitable for fan-type loads. The direction of rotation depends on whether the shading ring is located on the right or left side of the pole, so shaded pole motors are only suitable for uni-directional loads.

7

INVERTER-FED INDUCTION MOTOR DRIVES

INTRODUCTION

We saw in the previous chapter that the induction motor can only run efficiently at low slips, i.e. close to the synchronous speed of the rotating field. The best method of speed control must therefore provide for continuous smooth variation of the synchronous speed, which in turn calls for variation of the supply frequency. This is achieved using an inverter (as discussed in Chapter 2) to supply the motor. A complete speed control scheme which includes tacho (speed) feedback is shown in block diagram form in Figure 7.1.

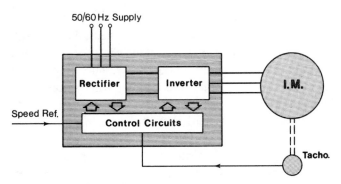

Figure 7.1 *General arrangement of inverter-fed variable-frequency induction motor controlled-speed drive*

Variable frequency inverter-fed induction motors are now used in ratings up to hundreds of kilowatts. Standard 50 Hz or 60 Hz motors are usually employed, and the inverter output frequency typically covers the range from around 5–10 Hz up to perhaps 120 Hz. This is sufficient to give at least a 10:1 speed range with top speed of twice the normal (mains frequency) operating speed. Most inverters require a 3-phase supply and provide a 3-phase output, but in the smaller sizes (say up to 5 kW) single-phase input versions are available. Some very small inverters (usually less than 1 kW) are intended for use with single-phase motors.

The vast majority of inverters are voltage source inverters (VSI), but current source inverters (CSI) are still used for some applications. Only the former will be discussed here.

Comparison with d.c. drive

D.C. drive technology has dominated the market for many years, and any challenger must compete with it in terms of price and performance. The majority of inverter-fed drives are therefore marketed as a package consisting of an inverter and a standard induction motor, and priced so as to be competitive with the thyristor d.c. drive of the same rating and speed range. The higher cost of the a.c. inverter compared with the thyristor d.c. converter is offset by the much lower cost of the standard induction motor compared with the d.c. motor.

There is a clear implication in much of the promotional material that the inverter-fed system can perform at least as well, or even better, than the d.c. drive. This is true for some applications such as fans and pumps where high torque is only needed at high speeds. But whereas a d.c. drive will invariably be supplied with a motor which is provided with through ventilation to allow it to operate continuously at low speeds without overheating, the standard induction motor has no such provision, having been designed primarily for fixed-frequency full-speed operation. Thus although the inverter may be capable of driving the induction motor with full torque at low speeds, continuous operation is unlikely to be possible because the motor will overheat. Suppliers of inverter drives are not anxious to emphasize this limitation, but users need to be aware of it.

The steady-state performance of inverter-fed drives is broadly comparable with that of d.c. drives (except for the limitation highlighted above), with drives of the same rating having similar overall efficiencies and overall torque-speed capabilities. Speed holding is likely to be less good in the induction motor drive, though if tacho feedback is used both systems will be excellent. The induction motor is clearly more robust and better suited to hazardous environments, and can run at very high speeds if necessary.

Some of the early inverters (e.g. from the early 1980s) did not employ pulse width modulation, and produced jerky rotation at low speed. They were also noticeably more noisy than their d.c. counterparts, but the widespread adoption of PWM has greatly improved these aspects. Some low and medium power inverters modulate at ultrasonic frequencies, which naturally results in much quieter operation.

The achilles heel of the inverter-fed system has been the relatively poor transient performance. For most applications this is not a serious drawback, but where rapid response to changes in speed or load is called for (e.g. in rolling mills), the basic d.c. drive with its fast-acting current control loop is inherently superior. It is possible to achieve equivalent levels of dynamic performance from induction motors, but until recently the complexity of the control needed was too great to be economically viable. However it is now possible to provide the sophisticated control by using the latest very large scale integrated circuits, and many manufacturers now offer this so-called 'vector' control (see later section) as an optional extra for high-performance drives.

Inverter-fed induction motors

When we looked at the converter-fed d.c. motor we saw that the behaviour was governed primarily by the mean d.c. voltage, and that for most purposes we could safely ignore the ripple components. A similar approximation is useful when looking at how the inverter-fed induction motor performs. We make use of the fact that although the actual voltage waveform supplied by the inverter will not be sinusoidal, the motor behaviour depends principally on the fundamental (sinusoidal) component of the applied voltage. This is a somewhat surprising but extremely welcome simplification, because it allows us to make use of our knowledge of how the induction motor behaves with a sinusoidal supply to anticipate how it will behave when fed from an inverter.

In essence, the reason why the harmonic components of the applied voltage are much less significant than the fundamental is that the impedance of the motor at the harmonic frequencies is much higher than at the fundamental frequency. This causes the current to be much more sinusoidal than the voltage, as shown in Figure 7.2, and this in turn means that we can expect a sinusoidal travelling field to be set up in much the same way as discussed in the previous chapter.

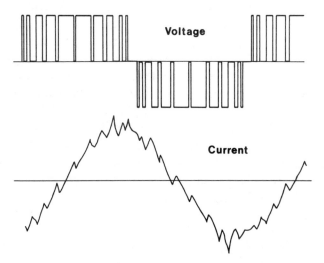

Figure 7.2 *Typical voltage and currrent waveforms for PWM inverter-fed induction motor*

It would be wrong to pretend that the harmonic components have no effects, of course. They can create unpleasant acoustic noise, and always give rise to additional iron and copper losses. As a result it is common for a standard motor to have to be de-rated (by up to perhaps 10 per cent) for use on an inverter supply.

Steady-state operation – importance of achieving full flux

Three simple relationships need to be borne in mind in order to simplify understanding of the inverter-fed characteristics. Firstly, we established in Chapter 5 that for a given induction motor, the torque developed depends on the strength of the rotating flux wave, and on the slip of the rotor, i.e. on the relative velocity of the rotor with respect to the flux wave. Secondly, the strength or amplitude of the flux wave depends directly on the supply voltage to the stator windings, and inversely on the supply frequency. And thirdly, the absolute speed of the flux wave depends directly on the supply frequency.

Recalling that the motor can only operate efficiently when the slip is small, we see that the basic method of speed control rests on the control of the speed of rotation of the flux wave (i.e the synchronous speed), by control of the supply frequency. If the motor is a 4-pole one, for example, the synchronous speed will be 1 500 rev/min when supplied at 50 Hz, 1 200 rev/min at 40 Hz, 750 rev/min at 25 Hz, and so on. The no-load speed will therefore be almost exactly proportional to the supply frequency, because the torque at no load is small and the corresponding slip is also very small.

Turning now to what happens on load, we know that when a load is applied the rotor slows down, the slip increases, more current is induced in the rotor, and more torque is produced. When the speed has reduced to the point where the motor torque equals the load torque, the speed becomes steady. We normally want the drop in speed with load to be as small as possible, so that the motor holds its speed in the face of load: in other words we want to minimize the slip for a given load.

We saw in Chapter 5 that the slip for a given torque depends on the amplitude of the rotating flux wave: the higher the flux, the smaller the slip needed for a given torque. It follows that having set the desired speed of rotation of the flux wave by controlling the output frequency of the inverter we must also ensure that the magnitude of the flux is adjusted so that it is at its full (rated) value, regardless of the speed of rotation. This is achieved by making the output voltage from the inverter vary in the appropriate way in relation to the frequency.

We recall that the amplitude of the flux wave is proportional to the supply voltage and inversely proportional to the frequency, so if we arrange that the voltage supplied by the inverter varies in direct proportion to the frequency, the flux wave will have a constant amplitude. This philosophy is at the heart of most inverter-fed drive systems. There are variations as we will see, but in the majority of cases the internal control of the inverter will be designed so that the output voltage to frequency ratio (V/f) is automatically kept constant, at least up to the 'base' (50 Hz or 60 Hz) frequency.

Most inverters are designed for direct connection to the mains supply, without a transformer, and as a result the maximum inverter output voltage is limited to a value similar to that of the mains. With a 415 V supply, for example, the maximum inverter output voltage will be perhaps 450 V. Since the inverter will normally be used to supply a standard induction motor designed for say 415 V, 50 Hz operation, it is obvious that when the inverter is set to deliver 50 Hz, the voltage should be 415 V, which is within the inverter's voltage range. But when the frequency was raised to say 100 Hz, the voltage should, ideally, be increased to 830 V in order to obtain full flux. The inverter cannot supply voltages above 450 V, and it follows that in this case full flux can only be maintained up to speeds a little above base speed. (It should be noted that even if the inverter could provide higher voltages, they could not be applied to a standard motor because the winding insulation will have been designed to withstand not more than the rated voltage.)

Established practice is for the inverter to be capable of maintaining the V/f ratio constant up to the base speed (50 Hz or 60 Hz), but to accept that at all higher frequencies the voltage will be constant at its maximum value. This means that the flux is maintained constant at speeds up to base speed, but beyond that the flux reduces inversely with frequency. Needless to say the performance above base speed is adversely affected, as we shall see.

Users are sometimes alarmed to discover that both voltage and frequency change when a new speed is demanded. Particular concern is expressed when the voltage is seen to reduce when a lower speed is called for. Surely, it is argued, it can't be right to operate say a 400V induction motor at anything less than 400V. The fallacy in this view should now be apparent: the figure of 400V is simply the correct voltage for the motor when run directly from the mains, at say 50 Hz. If this full voltage was applied when the frequency was reduced to say 25 Hz, the implication would be that the flux would have to rise to twice its rated value. This would greatly overload the magnetic circuit of the machine, giving rise to excessive saturation of the iron, an enormous magnetizing current, and wholly unacceptable iron and copper losses. To prevent this from happening, and keep the flux at its rated value, it is essential to reduce the voltage in proportion to frequency. In the case above, for example, the correct voltage at 25 Hz would be 200V.

TORQUE-SPEED CHARACTERISTICS – CONSTANT V/f OPERATION

If the voltage at each frequency is adjusted so that the ratio V/f is kept constant up to base speed, and full voltage is applied thereafter, a family of torque speed curves as shown in Figure 7.3 is obtained.

These curves are typical for a standard induction motor of several kW output.

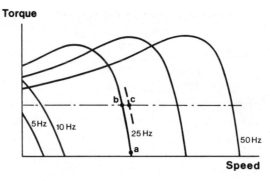

Figure 7.3   *Torque-speed curves for inverter-fed induction motor with constant voltage/frequency ratio*

As expected, the no-load speeds are directly proportional to the frequency, and if the frequency is held constant e.g at 25 Hz in Figure 7.3, the speed drops only modestly from no-load (point a) to full-load (point b). These are therefore good open-loop characteristics, because the speed is held fairly well from no-load to full-load. If the application calls for the speed to be held precisely, this can clearly be achieved (with the aid of closed-loop speed control) by raising the frequency so that the full-load operating point moves to point (c).

We note also that the pull-out torque and the torque stiffness (i.e. the slope of the torque-speed curve in the normal operating region) is more or less the same at all points below base speed, except at low frequencies where the effect of stator resistance in reducing the flux becomes very pronounced. It is clear from Figure 7.3 that the starting torque at the minimum frequency is much less than the pull-out torque at higher frequencies, and this could be a problem for loads which require a high starting torque.

The low-frequency performance can be improved by increasing the V/f ratio at low frequencies, a technique which is referred to as 'low-speed voltage boosting'. Most drives incorporate provision for some form of voltage boost, either by way of a single adjustment to allow the user to set the

Plate 7.1   *PWM variable-frequency inverter drive for high-speed machine tool applications (Photograph by courtesy of GEC Industrial Controls Ltd)*

desired starting torque, or by means of more complex provision for varying the V/f ratio over a range of frequencies. A typical set of torque-speed curves for a drive with voltage boost is shown in Figure 7.4.

The curves in Figure 7.4 have an obvious appeal because they indicate that the motor is capable of producing prac-

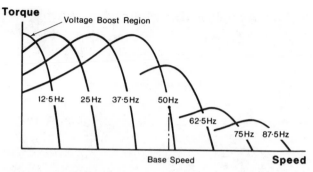

Figure 7.4   *Typical torque-speed curves for inverter-fed induction motor with low-speed voltage boost, constant voltage/frequency ratio from low speed up to base speed, and constant voltage above base speed*

tically the same maximum torque at all speeds up to the base (50 Hz or 60 Hz) speed. This region of the characteristics is known as the 'constant torque' region, which means that for frequencies up to base speed, the maximum possible torque which the motor can deliver is independent of the set speed. Continuous operation at peak torque will not be allowable because the motor will overheat, so an upper limit will be imposed by the controller, as discussed shortly. With this imposed limit, operation below base speed corresponds to the armature-voltage control region of a d.c. drive.

Beyond the base frequency, the V/f ratio reduces because V remains constant. The amplitude of the flux wave therefore reduces inversely with the frequency. Now we saw in Chapter 5 that the pull-out torque always occurs at the same absolute value of slip of the rotor relative to the travelling field, and that the peak torque is proportional to the square of the flux density. Hence in the constant-voltage region the peak torque reduces inversely with the square of the frequency. This rapid fall-off is indicated in Figure 7.4.

Permissible range of continuous operation

Although the curves in Figure 7.4 show what torque the motor can produce for each frequency and speed, they give no indication of whether continuous operation is possible at each point, yet this matter is of course extremely important from the user's viewpoint. Unfortunately there is as yet no established norm for inverter-fed drives to match that for d.c drives. Hence whilst the d.c. drive customer can safely assume that he can operate continuously at rated torque at all speeds up to base speed, and at full power in the field weakening region, no such simple convention can be relied on for inverter-fed drives.

The reasons for the differences between d.c. drives and inverter-fed drives stem firstly from different philosophies in relation to the motor, and secondly from the fact that whereas d.c. converters represent a mature and stable technology, inverters are still evolving and no universally accepted practice has yet been established.

Limitations imposed by the inverter – constant power and constant torque regions

The main concern in the inverter is to limit the currents to a safe value as far as the main switching devices are concerned. The current limit will be at least equal to the rated current of the motor, and the inverter control circuits will be arranged so that no matter what the user does the output current cannot exceed a safe value.

The current limit feature imposes an upper limit on the permissible torque in the region below base speed. This will normally correspond to the rated torque of the motor, which is typically about half the pull-out torque, as indicated by the shaded region in Figure 7.5.

In the region below base speed, the motor can therefore develop any torque up to rated value at any speed (but not necessarily for prolonged periods, as discussed below). This region is therefore known as the 'constant torque' region, and it corresponds to the armature voltage control region of a d.c. drive.

Above base speed the imposition of a stator current limit results in the maximum permissible torque reducing inversely with the speed, as shown in Figure 7.5. This region is

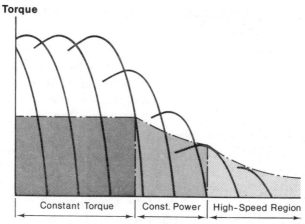

Torque

Constant Torque | Const. Power | High-Speed Region

Figure 7.5   *Constant torque, constant power and high-speed motoring regions*

therefore known as the 'constant power' region. There is of course a close parallel with the d.c. drive here, both systems operating with reduced or weak field in the constant power region. The region of constant power normally extends to somewhere around twice base speed, and the motor is allowed to operate with higher slips than below base speed. The rotor currents are not excessive, however, because the flux is reduced.

At the upper limit of the constant power region, the current limit coincides with the pull-out torque limit. Operation at still higher speeds is sometimes provided, but constant power is no longer available because the maximum torque is limited to the pull-out value, which reduces inversely with the square of the frequency. In this high-speed motoring region (Figure 7.5) the limiting torque-speed relationship is similar to that of a series d.c. motor.

Limitations imposed by motor

The standard practice in d.c. drives is to use a motor specifically designed for operation from a thyristor converter. The motor will have a laminated frame, will probably come complete with a tachogenerator, and – most important of all – will have been designed for through ventilation and equipped with an auxiliary air blower. Adequate ventilation is guaranteed at all speeds, and continuous operation with full torque (i.e. full current) at even the lowest speed is therefore in order.

By contrast, the current practice in inverter-fed systems is for a standard industrial induction motor to be supplied with the inverter. These motors are totally enclosed, with an external shaft-mounted fan which blows air over the finned outer case. They are designed first and foremost for continuous operation from the fixed frequency mains, and running at base speed.

When such a motor is operated at a low frequency (e.g. 10 Hz), the speed is much lower than base speed and the efficiency of the cooling fan is greatly reduced. At the lower speed the motor will be able to produce as much torque as at base speed (see Figure 7.5) but in doing so the losses in both stator and rotor will also be more or less the same as at base speed. Since the fan was only just adequate to prevent overheating at base speed, it is inevitable that the motor will overheat if operated at full torque and low speed for any length of time. Manufacturers and suppliers are understandably not keen to emphasize this limitation, and no well-established pattern has yet emerged. Users therefore need to raise the question with the supplier before committing themselves. The real problem lies in the use of a standard motor; when through-ventilated motors with integral blowers become the accepted standard, the inverter-fed system will be freed of most of its current limitations.

One recent trend designed to mitigate against the danger of motor overheating at low speeds is for inverter suppliers to design their control circuits so that the flux and current limit are deliberately reduced at low speeds. The constant-torque facility is thus sacrificed in order to reduce copper and iron losses, but as a result the drive is only suitable for fan or pump type loads which do not require high torque at low speed. These systems inevitably compare badly with d.c. drives, though they manage to save face by being promoted as 'energy-saving' drives.

CONTROL ARRANGEMENTS FOR INVERTER-FED DRIVES

Because inverter-fed drives are relatively new, no one particular approach has yet emerged as the best control philosophy in the way that it has for the d.c. drive (Chapter 4). For speed control manufacturers offer options ranging in sophistication from a basic open-loop scheme which is adequate when precise speed holding is not essential, through closed-loop schemes with tacho or encoder feedback, up to vector control schemes which are necessary when optimum dynamic performance is called for. We will therefore look briefly at each type.

Open-loop speed control

In the smaller sizes the simple 'constant V/f' control is the most popular, and is shown in Figure 7.6.

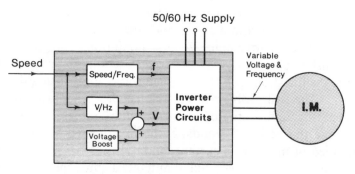

Figure 7.6 *Schematic diagram of open-loop inverter-fed induction motor speed controlled drive*

The output frequency, and hence the no-load speed of the motor, is set by the speed reference signal, which is usually an analogue voltage (0–10 V) or current (4–20 mA). This set-speed signal can be obtained either from a potentiometer on the front panel, or remotely from elsewhere. Some modest adjustment of the V/f ratio may be possible, and the user can normally adjust the low-speed voltage boost, the maximum and minimum speeds, and the acceleration and deceleration rates.

Typical steady-state operating torque-speed curves are shown in Figure 7.7.

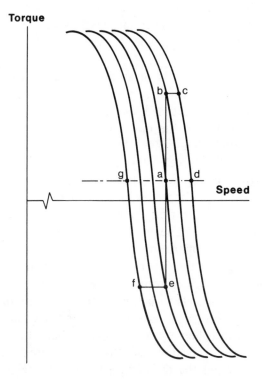

Figure 7.7   *Acceleration and deceleration trajectories in the torque-speed plane*

For each set speed (i.e each frequency) the speed remains reasonably constant because of the stiff torque-slip characteristic of the cage motor. If the load is increased beyond rated torque, the internal current limit comes into play to prevent the motor from reaching the unstable region beyond pull-out. Instead, the frequency and speed are reduced, so that the system behaves in the same way as a d.c. drive.

Sudden changes in the speed reference are buffered by the action of an internal frequency ramp signal, which causes the frequency to be gradually increased or decreased. If the load inertia is low, the acceleration will be accomplished without the motor entering the current-limit regime. On the other hand if the inertia is large, the acceleration will take place along the torque-speed trajectory shown in Figure 7.7.

Suppose the motor is originally operating in the steady-state at point (a), when a new higher speed is demanded. The frequency is increased, causing the motor torque to rise to point (b), where the current has reached the allowable limit. The rate of increase of frequency is then reduced so that the motor accelerates under constant-current conditions to point (c), where the current falls below the limit and operation finally settles at point (d).

A typical deceleration trajectory is shown by the path aefg in Figure 7.7. The torque is negative for much of the time, the motor operating in quadrant 2 and regenerating kinetic energy to the inverter. Most small inverters do not have the capability to return power to the a.c. supply, and the excess energy has to be dissipated in a resistor inside the converter. The resistor is usually connected across the d.c. link, and controlled by a chopper. When the link voltage tends to rise, because of the regenerated energy, the chopper switches the resistor on to absorb the energy. High inertia loads which are subjected to frequent deceleration can therefore pose problems of excessive power dissipation in the dump resistor.

Speed reversal poses no problem, the inverter firing sequence being reversed automatically at zero speed, thereby allowing the motor to proceed smoothly into quadrants 3 and 4.

Closed-loop speed control

Where precision speed holding is required a closed-loop scheme is used, with speed feedback from either a d.c. or a.c. tachogenerator, or a digital shaft encoder. A typical arrangement is shown in Figure 7.8.

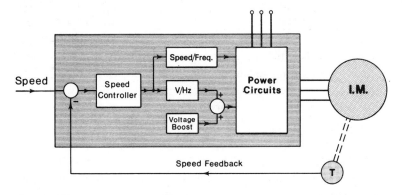

Figure 7.8 *Schematic diagram of closed-loop inverter-fed induction motor speed controlled drive with tacho feedback*

Analogue control using a proportional and integral speed error amplifier (see also the d.c. drive, Chapter 4) can give a good transient response with steady-state speed holding of better than 1 per cent for a speed range of 20:1 or more. For higher precision, a shaft encoder together with a phase-locked loop is used.

The need to fit a tacho or encoder can be a problem if a standard induction motor is used, because there is normally no shaft extension at the non-drive end. The user then faces the prospect of paying a great deal more for what amounts to a relatively minor modification, simply because the motor then ceases to be standard.

Vector control

As mentioned earlier, simple inverter-fed drives compare unfavourably with d.c. drives as far as their transient or dynamic performance is concerned. In essence this is because the torque in a d.c. motor is directly proportional to the armature current, which can be directly controlled even under transient conditions, whereas the induction motor transient torque depends in a complex way on the behaviour of the air-gap flux and the induced rotor currents.

For the majority of applications a simple inverter-fed scheme is nevertheless perfectly adequate, but for some very demanding tasks, such as high-speed machine tool spindle drives, the dynamic performance is extremely important and 'vector' or 'field-oriented' control is warranted. This approach has been under development for many years, but has only recently begun to be offered as an option by an increasing number of drive manufacturers.

Understanding all the ins and outs of vector control is well beyond our scope, and is anyway unnecessary from the user viewpoint. In essence, however, vector control allows both the magnitude and the instantaneous position of the airgap flux wave relative to the rotor current wave to be controlled so that at every instant the torque is maximized. This compares with an ordinary or 'scalar' control in which only the steady-state magnitude and speed of the flux wave can be controlled. Because the induction motor is a complex multi-variable non-linear system, vector control requires a large number of fast computations to be continually carried out so that the right instantaneous voltages are applied to each stator winding. This has only recently been made possible with the use of VLSI circuitry as part of the drive control.

As far as the user is concerned two points need to be stressed. Firstly, the steady-state performance is not improved by vector control, so unless special dynamic performance is called for, vector control offers no advantage. And secondly, users should be aware that many vector control schemes rely on signals from a shaft encoder, which precludes the use of a standard motor.

8

STEPPING MOTOR SYSTEMS

INTRODUCTION

Stepping motors have become very popular because they can
be controlled directly by computers, microprocessors or
programmable controllers. Their unique feature is that the
output shaft rotates in a series of discrete angular intervals,
or steps, one step being taken each time a command pulse is
received. When a definite number of pulses has been
supplied, the shaft will have turned through a known angle,
and this makes the motor ideally suited for open-loop position control.

The idea of a shaft progressing in a series of steps could
easily conjure up visions of a ponderous device laboriously
indexing until the target number of steps has been reached,
but this would be quite wrong. Each step is completed very
quickly, usually in a few milliseconds; and when a large
number of steps is called for the step command pulses can be
delivered rapidly, sometimes as fast as several thousand
steps per second. At these high stepping rates the shaft rotation becomes smooth, and the behaviour resembles that of
an ordinary motor. Typical applications include floppy-disc
head drives, and small numerically-controlled machine
tool slides, where the motor would drive a lead screw; and
daisy-wheel print heads, where the motor might drive the
head directly, or via a belt.

Most stepping motors look very much like conventional motors, and as a general guide we can assume that the torque and power of a stepping motor will be similar to the torque and power of a conventional motor of the same dimensions. Step angles are mostly in the range 1.8°–90°, with torque ranging from 1 μNm (in a tiny wristwatch motor of 3 mm diameter) up to perhaps 40 Nm in a motor of 15 cm diameter suitable for a machine tool application where speeds of 500 rev/min might be called for. The majority of applications fall between these limits, and use motors which can comfortably be held in the hand.

Open-loop position control

A basic stepping motor system is shown in Figure 8.1.

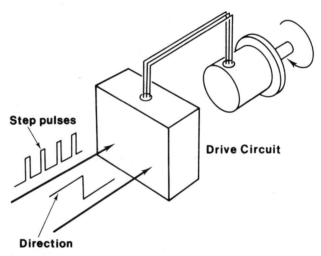

Figure 8.1   *Open-loop position control using a stepping motor*

The drive circuit contains the electronic switching circuits which supply the motor, and is discussed later. The output is the angular position of the motor shaft, while the input consists of two low-power digital signals. Every time a pulse occurs on the step input line, the motor takes one step, the shaft remaining at its new position until the next step pulse is supplied. The state of the direction line ('high' or 'low') determines whether the motor steps clockwise or anticlockwise. A given number of step pulses will therefore cause the output shaft to rotate through a definite angle.

This one to one correspondence between pulses and steps is the great attraction of the stepping motor: it provides *position* control, because the output is the angular position of the output shaft. It is a *digital* system, because the total angle turned through is determined by the *number* of pulses supplied; and it is *open-loop* because no feedback need be taken from the output shaft.

Generation of step pulses and motor response

The step pulses are usually produced by an oscillator circuit, which is itself controlled by an analogue voltage, digital controller or microprocessor. When a given number of steps is to be taken, the oscillator is started, and the pulses generated are counted, until the required number of steps is reached, when the oscillator is gated off. This is illustrated in Figure 8.2, for a five step sequence. There are five step command pulses, equally spaced in time, and the motor takes one step following each pulse.

Three important general features can be identified with reference to Figure 8.2. Firstly, although the total angle turned through (5 steps) is governed only by the number of pulses, the average speed of the shaft (which is shown by the slope of the broken line in Figure 8.2) depends on the oscillator frequency. The higher the frequency, the shorter the time taken to complete the five steps.

Secondly, the stepping action is not perfect. The rotor takes a finite time to move from one position to the next, and then overshoots and oscillates before finally coming to rest at the new position. Overall single-step times vary with motor size, step angle and the nature of the load, but are commonly within the range 5–100 ms. This is often fast enough not to be seen by the unwary newcomer, though

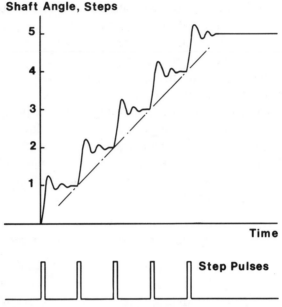

Figure 8.2   *Typical step response to low-frequency train of step command pulses*

individual steps can usually be heard; small motors 'tick' when they step, and larger ones make a satisfying 'click' or 'clunk'.

Thirdly, in order to be sure of the absolute position at the end of a stepping sequence, we must know the absolute position at the beginning. This is because a stepping motor is an incremental device. As long as it is not abused, it will always take one step when a drive pulse is supplied, but in order to keep track of absolute position simply by counting the number of drive pulses (and this is after all the main virtue of the system) we must always start the count from a known datum position. Normally the step counter will be 'zeroed' with the motor shaft at the datum position, and will then count up for clockwise direction, and down for anticlockwise rotation. Provided no steps are 'lost' (see later) the number in the step counter will then always indicate the absolute position.

High speed running and ramping

The discussion so far has been restricted to operation when the step command pulses are supplied at a constant rate, and with sufficiently long intervals between the pulses to allow the rotor to come to rest between steps. Very large numbers of small stepping motors in watches and clocks do operate continuously in this way, stepping once every second, but most commercial and industrial applications call for a more exacting and varied performance.

To illustrate the variety of operations which might be involved, and to introduce high-speed running, we can look briefly at a typical industrial application. A stepping motor-driven table feed on a numerically-controlled milling machine nicely illustrates both of the key operational features discussed earlier. These are the ability to control position (by supplying the desired number of steps) and velocity (by controlling the stepping rate).

The arrangement is shown diagrammatically in Figure 8.3. The motor is coupled to a leadscrew on the worktable, so that each motor step causes a precise incremental movement of the workpiece relative to the cutting tool. By making the increment small enough, the fact that the motion is discrete rather than continuous will not cause any difficulties in the machining process. We will assume that we have selected the step angle, the pitch of the leadscrew, and any necessary gearing so as to give a table movement of 0.01 mm per motor step. We will also assume that the necessary step command

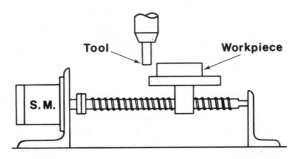

Figure 8.3   *Application of stepping motor for open-loop position control*

pulses will be generated by a digital controller or computer, programmed to supply the right number of pulses, at the right speed for the work in hand.

If the machine is a general-purpose one, many different operations will be required. When taking heavy cuts, or working with hard material, the work will have to be offered to the cutting tool slowly, at say, 0.02 mm/sec. The stepping rate will then have to be set to 2 steps/sec. If we wish to mill out a slot 10 cms long, we will therefore programme the controller to put out 10 000 steps, at a uniform rate of 2 steps/sec., and then stop. On the other hand, the cutting speed in softer material could be much higher, with stepping rates in the range 10–100 steps/second being in order. At the completion of a cut, it will be necessary to traverse the work back to its original position, before starting another cut. This operation needs to be done as quickly as possible, to minimize unproductive time, and a stepping rate of perhaps 2 000 steps/second (or even higher), may be called for.

It was mentioned earlier that a single step (from rest) takes upwards of several milliseconds. It should therefore be clear that if the motor is to run at 2 000 steps/sec. (i.e. 0.5 ms/step), it cannot possibly come to rest between successive steps, as it does at low stepping rates. Instead, we find in practice that at these high stepping rates, the rotor velocity becomes quite smooth, with hardly any outward hint of its stepwise origins.

Nevertheless, the vital one-to-one correspondence between step command pulses and steps taken by the motor is maintained throughout, and the open-loop position control feature is preserved. This extraordinary ability to operate at very high stepping rates (up to 20 000 steps/second in some motors), and yet to remain fully in synchronism with the command pulses, is the most striking feature of stepping motor systems.

Operation at high speeds is referred to as 'slewing'. The transition from single-stepping (as shown in Figure 8.2) to high-speed slewing is a gradual one and is indicated by the sketches in Figure 8.4. Roughly speaking, the motor will 'slew' if its stepping rate is above the frequency of its single-step oscillations. When motors are in the slewing range, they generally emit an audible whine, with a fundamental frequency equal to the stepping rate.

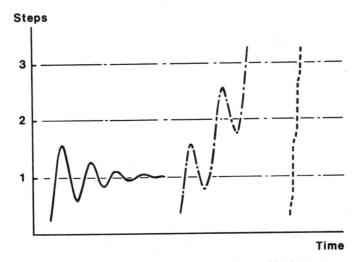

Figure 8.4   *Position-time responses at low, medium and high stepping rates*

It will come as no surprise to learn that a motor cannot be started from rest and expected to 'lock on' directly to a train of command pulses at, say, 2000 steps/second, which is well into the slewing range. Instead, it has to be started at a more modest stepping rate, before being accelerated (or 'ramped') up to speed. In undemanding applications, the ramping can be done slowly, and spread over a large number of steps; but if the high stepping rate has to be reached quickly, the timings of individual step pulses must be very precise.

We may wonder what will happen if the stepping rate is increased too quickly. The answer is simply that the motor will not be able to remain 'in step' and will stall. The step command pulses will still be being delivered, and the step counter will be accumulating what it believes are motor steps, but, by then, the system will have failed completely. A similar failure mode will occur if, when the motor is slewing, the train of step pulse is suddenly stopped, instead of being progressively slowed. The stored kinetic energy of the motor (and load) will cause it to overrun, so that the number of motor steps will be greater than the number of command pulses. Failures of this sort are prevented by the use of closed-loop control, as discussed later.

PRINCIPLE OF MOTOR OPERATION

The principle on which stepping motors are based is very simple. When a bar of iron or steel is suspended so that it is free to rotate in a magnetic field, it will align itself with the field. If the direction of the field is changed, the bar will turn until it is again aligned, by the action of the so-called reluctance torque.

The two most important types of motor (judged on the basis of their share of the total market) are the variable-reluctance (VR) type and the hybrid type. Both types utilize the reluctance principle, the difference between them lying in the method by which the magnetic fields are produced. In the VR type the fields are produced solely by sets of stationary current-carrying windings. The hybrid type also has sets of windings, but the addition of a permanent magnet (on the rotor) gives rise to the description 'hybrid' for this type of motor. Although both types of motor work on the same basic principle, it turns out in practice that the VR type is attractive for the larger step angles (eg. 15°, 30°, 45°), while the hybrid tends to be best-suited when small angles (e.g. 1.8°, 2.5°) are required.

Variable reluctance motor

A simplified diagram of a 30°/step VR stepping motor is shown in Figure 8.5. The stator is made from a stack of steel laminations, and has six equally-spaced projecting poles, or teeth, each carrying a separate simple coil. The rotor, which may be solid or laminated, has four projecting teeth, of the same width as the stator teeth. There is a very small air gap – typically between 0.02 mm and 0.2 mm – between rotor and stator teeth. When no current is flowing in any of the stator coils, the rotor will therefore be completely free to rotate. Diametrically opposite pairs of stator coils are connected in series, such that when one of them acts as a north pole, the other acts as a south pole. There are thus three independent stator circuits, or phases, and each one can be supplied with direct current from the drive circuit (not shown in Figure 8.5).

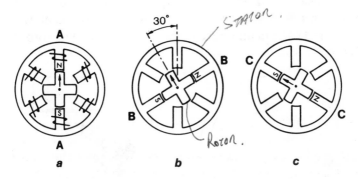

a   *b*   *c*

Figure 8.5   *Principle of operation of 30°/step variable-reluctance stepping motor*

When phase A is energized, a magnetic field with its axis along the stator poles of phase A is created. The rotor is therefore attracted into a position where a pair of (diametrically opposite) rotor poles line up with the field, i.e. in line with the phase-A pole, as shown in Figure 8.5(a). When phase A is switched off, and phase B is switched on instead, the second pair of rotor poles will be pulled into alignment with the stator poles of phase B, the rotor moving through 30° anticlockwise to its new step position, as shown in Figure 8.5(b). A further anticlockwise step of 30° will occur when phase B is switched off and phase C is switched on. At this stage the original pair of rotor poles come into play again, but this time they are attracted to stator poles C, as shown in Figure 8.5(c). By repetitively switching on the stator phases in the sequence ABCA etc. the rotor will rotate anticlockwise in 30° steps, while if the sequence is ACBA etc. it will rotate clockwise. This mode of operation is known as '1-phase-on', and is the simplest way of making the motor step. Note that the polarity of the energizing current is not significant. The motor will be aligned equally well regardless of the direction of current.

Hybrid motor

A cross-sectional view of a typical 1.8° hybrid motor is shown in Figure 8.6. The stator has 8 main poles, each with 5 teeth, and each main pole carries a simple coil. The rotor has

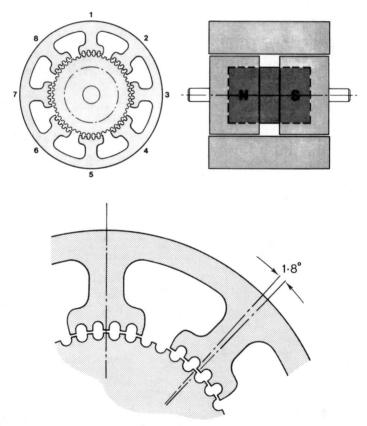

Figure 8.6   *Hybrid (200 step/rev) stepping motor. The detail shows the rotor and stator tooth alignments, and indicates the step angle of 1.8°*

two end-caps, each with 50 teeth, and separated by a permanent magnet. The rotor teeth have the same pitch as the teeth on the stator poles, and are offset so that the centre line of a tooth at one end coincides with a slot at the other. The permanent-magnet is axially magnetised, so that one set of rotor teeth is given a north polarity, and the other a south polarity.

When no current is flowing in the windings, the only source of magnetic flux across the air-gap is the permanent magnet. If there was no offset between the two sets of rotor teeth, there would be a strong periodic alignment torque when the rotor was turned, and every time a set of stator teeth was in line with the rotor teeth we would obtain a stable equilibrium position. However there is an offset, and this causes the alignment torque due to the magnet to be almost eliminated. In practice a small 'detent' torque remains, and this can be felt if the shaft is turned when the motor is de-energized: the motor tends to be held in its step positions by the detent torque. This is sometimes very useful: for example it is usually enough to hold the rotor stationary when the power is switched off, so the motor can be left overnight without fear of it being accidentally moved to a new position.

The 8 coils are connected to form two phase-windings. The coils on poles 1, 3, 5 and 7 form phase A, while those on 2, 4, 6 and 8 form phase B. When phase A carries positive current stator poles 1 and 5 are magnetized as south, and poles 3 and 7 become north. The teeth on the north end of the rotor are attracted to poles 1 and 5 while the offset teeth at the south end of the rotor are attracted into line with the teeth on poles 3 and 7. To make the rotor step, phase A is switched off, and phase B is energized with either positive current or negative current, depending on the sense of rotation required. This will cause the rotor to move by one quarter of a tooth pitch (1.8°) to a new equilibrium (step) position.

The motor is continuously stepped by energizing the phases in the sequence +A, −B, −A, +B, +A (clockwise) or +A, +B, −A, −B, +A (anticlockwise). It will be clear from this that a bi-polar supply is needed (i.e. one which can furnish +ve or −ve current). When the motor is operated in this way it is referred to as '2-phase, with bi-polar supply'.

If a bi-polar supply is not available, the same pattern of pole energization is achieved in a different way. Each pole carries two identical ('bifilar wound') coils. To magnetize pole 1 north, a positive current is fed into one set of phase-A coils. But to magnetize pole 1 south, the same positive current is fed into the other set of phase-A coils, which have the opposite winding sense. In total, there are then four separate windings, and when the motor is operated in this way it is referred to as '4-phase, with uni-polar supply'. Since only half of the winding is used at any one time, the thermally-rated output is clearly reduced as compared with bi-polar operation.

Summary

The construction of stepping motors is simple, the only moving part being the rotor, which has no windings,

Plate 8.1   *1.8° (200 step/rev) hybrid stepping motors. The construction of this type of motor is shown in Figure 8.6. Where more torque is called for the stator is extended to accommodate two or three rotor 'stack' assemblies. (Photograph by courtesy of Evershed and Vignoles Ltd)*

commutator or brushes: they are therefore robust and reliable. The rotor is held at its step position solely by the action of the magnetic flux between stator and rotor. The step angle is a property of the tooth geometry and the arrangement of the stator windings, and accurate punching and assembly of the stator and rotor laminations is therefore necessary to ensure that adjacent step position are exactly equally spaced. Any errors due to inaccurate punching will be non-cumulative, however.

The step angle is obtained from the expression

$$\text{Step Angle} = \frac{360°}{(\text{Rotor Teeth}) \times (\text{Stator Phases})}$$

The VR motor in Figure 8.5 has 4 rotor teeth, 3 stator phase-windings, and the step angle is therefore 30°, as already shown. It should also be clear from the equation why small angle motors always have to have a large number of rotor teeth. Probably the most widely used motor is the 200 step/rev hybrid type (see Figure 8.6). This has a 50-tooth rotor, 4-phase stator, and hence a step angle of 1.8°

$$\left(= \frac{360°}{50 \times 4} \right).$$

The magnitude of the aligning torque clearly depends on the magnitude of the current in the phase-winding. However, the equilibrium position itself does not depend on the magnitude of the current, because it is simply the position where the rotor and stator teeth are in line. This property underlines the digital nature of the stepping motor.

MOTOR CHARACTERISTICS

Static torque-displacement curves

From the previous discussion, it should be clear that the shape of the torque-displacement curve, and in particular the peak static torque, will depend on the internal electromagnetic design of the rotor. In particular the shapes of the rotor and stator teeth, and the disposition of the stator windings (and permanent magnet(s)) all have to be optimized to obtain the maximum static torque. We now turn to a typical static torque-displacement curve, and look at how it determines motor behaviour. Several aspects will be discussed, including the explanation of basic stepping (which has already been looked at in a qualitative way), the influence of load torque on step position accuracy, the effect of the amplitude of the winding current, and half-step and mini-stepping operation. For the sake of simplicity, the discussion will be based on the 30°/step 3-phase VR motor introduced earlier, but the conclusions reached apply to any stepping motor.

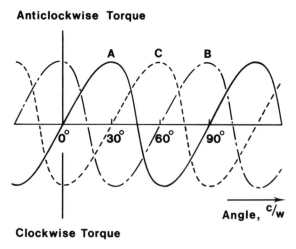

Figure 8.7   *Static torque-displacement curves for 30°/step variable reluctance stepping motor*

Typical torque-displacement curves for a 3-phase 30°/step VR motor are shown in Figure 8.7. There are 3 curves, one for each of the three phases, and for each curve we assume that the relevant phase winding carries its full (rated) current. The convention used in Figure 8.7 is that a clockwise displacement of the rotor corresponds to a movement to the right, while a positive torque tends to move the rotor anticlockwise.

When only one phase, say A, is energized, the other two phases exert no torque, so their curves can be ignored and we can focus attention on the solid line in Figure 8.7. Stable equilibrium positions (for phase A excited) exist at at $\theta = 0°$, 90°, 180° and 270°. They are stable (step) positions because any attempt to move the rotor away from them is resisted by a counteracting or restoring torque. These points correspond to positions where successive rotor poles (which are 90° apart) are aligned with the stator poles of phase A, as shown in Figure 8.5(a). There are also four unstable equilibrium positions, (at $\theta = 45°$, 135°, 225° and 315°) at which the torque is also zero. These correspond to rotor positions where the stator poles are mid-way between two rotor poles, and they are unstable because if the rotor is deflected in either direction from an unstable position, it will be accelerated in the same direction until it reaches the next stable position. If the rotor is free to turn, it will therefore always settle in one of the four stable positions.

Single-stepping

If we assume that phase A is energized, and the rotor is at rest in the position $\theta = 0°$ (Figure 8.7) we know that if we want to step in a clockwise direction, the phases must be energized in the sequence ACBA etc., so we can now imagine that phase A is switched off, and phase C is energized instead. We will also assume that the current is instantaneously transferred from phase A to phase C.

The rotor will find itself at $\theta = 0°$, but it will now experience a clockwise torque (see Figure 8.7) produced by phase C. The rotor will therefore accelerate clockwise, and will continue to experience clockwise torque, until it has turned through 30°. The rotor will be accelerating all the time, and it will therefore overshoot the 30° position, which is of course its target (step) position for phase C. As soon as it overshoots, however, the torque reverses, and the rotor experiences a braking torque, which brings it to rest before accelerating it back towards the 30° position. If there was no friction or other cause of damping, the rotor would of course continue to oscillate; but in practice it comes to rest at its new position quite quickly in much the same way as a damped second-order system. The next 30° step is achieved in the same way, by switching off the current in phase C, and switching on phase B.

In the above discussion, we have recognized that the rotor is acted on sequentially by each of the three separate torque curves shown in Figure 8.7. Alternatively, since the three curves have the same shape, we can think of the rotor being influenced by a single torque curve which 'jumps' by one step (30° in this case) each time the current is switched from one phase to the next. This is often the most convenient way of visualizing what is happening in the motor.

Step position error and holding torque

In the previous discussion the load torque was assumed to be zero, and the rotor was therefore able to come to rest with its poles exactly in line with the excited stator poles. When load torque is present, however, the rotor will not be able to pull fully into alignment, and a step position error will be unavoidable.

The origin and extent of the step position error can be seen in the typical torque-displacement curve shown in Figure 8.8. The true step position is indicated at 0 in the figure, and this is where the rotor would come to rest in the absence of load torque. If we imagine the rotor is initially at this position, and then consider that a clockwise load (T_L) is applied, the rotor will move clockwise, as it does so it will develop progressively more anticlockwise torque. The equilibrium position will be reached when the motor torque is equal and opposite to the load torque, i.e. at point A in Figure 8.8. The corresponding angular displacement from the step position (θ_e in Figure 8.8) is the step position error.

Static Torque

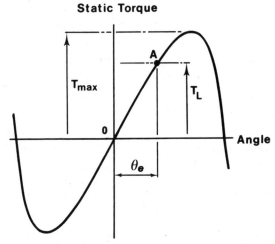

Figure 8.8 *Static torque-angle curve showing origin of step position error*

The existence of a step position error is one of the drawbacks of the stepping motor. The motor designer attempts to combat the problem by aiming to produce a steep torque-angle curve around the step position, and the user has to be aware of the problem and choose a motor with a sufficiently steep curve to keep the error within acceptable limits. In some cases this may mean selecting a motor with a higher peak torque than would otherwise be necessary, simply to obtain a steep enough torque curve around the step position.

As long as the load torque is less than T_{max} (see Figure 8.8), a stable rest position is obtained, but if the load torque exceeds T_{max}, the rotor will be unable to hold its step position. T_{max} is therefore known as the 'holding' torque. The value of the holding torque immediately conveys an idea of the overall capability of any motor, and it is – after step angle – the most important single parameter which is looked for in selecting a motor. Often, the adjective 'holding' is dropped altogether: for example '1 Nm motor' is understood to be one with a peak static torque (holding torque) of 1 Nm.

Half stepping

We have already seen how to step the motor in 30° increments by energizing the phases one at a time in the sequence ABCA etc. Although this '1-phase-on' mode is the simplest and most widely used, there are two other modes which are also frequently used. These are referred to as the '2-phase-on' mode and the 'half-stepping' mode. The 2-phase-on can provide greater holding torque and a much better damped single-step response than the 1-phase-on mode; and the half stepping mode permits the effective step angle to be halved – thereby doubling the resolution – and produces a smoother shaft rotation.

In the 2-phase-on mode, two phases are excited simultaneously. When phases A and B are energized, for example, the rotor experiences torques from both phases, and comes to rest at a point midway between the two adjacent full step positions. If the phases are switched in the sequence AB, BC, CA, AB etc., the motor will take full (30°) steps, as in the 1-phase-on mode, but its equilibrium positions will be interleaved between the full-step positions.

To obtain 'half-stepping' the phases are excited in the sequence A, AB, B, BC etc., i.e. alternately in the 1-phase-on and 2-phase-on modes. This is sometimes known as 'wave' excitation, and it causes the rotor to advance in steps of 15°, or half the full step angle. As might be expected, continuous half-stepping usually produces a smoother shaft rotation than full-stepping, and it also doubles the resolution

We can see what the static torque curve looks like when two phases are excited by graphical addition of the two separate phase curves. An example is shown in Figure 8.9, from which it can be seen that for this machine, the holding torque (i.e. the peak static torque) is higher with two phases excited than with only one excited. The stable equilibrium (half-step) position is at 15°, as expected. The price to be paid for the increased holding torque is the increased power dissipation in the windings, which is doubled as compared with the 1-phase-on mode. The holding torque increases by a

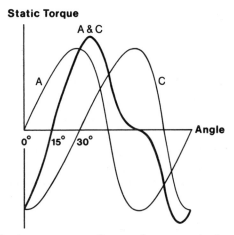

Static Torque

Figure 8.9   *Static torque curves for two-phase-on excitation*

factor less than two, so the torque per watt (which is a useful figure of merit) is reduced.

A word of caution is needed in regard to the adding-up of the two separate 1-phase-on torque curves to obtain the 2-phase-on curve. Strictly, such a procedure is only valid where the two phases are magnetically independent, or the common parts of the magnetic circuits are unsaturated. This is not the case in most motors, in which the phases share a common magnetic circuit which operates under highly saturated conditions. Direct addition of the 1-phase-on curve cannot therefore be expected to give an accurate result for the 2-phase-on curve, but it is easy to do, and provides a reasonable estimate.

Apart from the higher holding torque in the 2-phase-on mode, there is another important difference which distinguishes the static behaviour from that of the 1-phase-on mode. In the 1-phase-on mode, the equilibrium or step positions are determined solely by the geometry of the rotor and stator; they are the positions where the rotor and stator are in line. In the 2-phase-on mode, however, the rotor is intended to come to rest at points where the rotor poles are lined-up midway between the stator poles. This position is not sharply defined by the 'edges' of opposing poles, as in the 1-phase-on case; and the rest position will only be exactly midway if (a) there is exact geometrical symmetry and, more importantly (b) the two currents are identical. If one of the phase currents is larger than the other, the rotor will come to rest closer to the phase with the higher current, instead of half-way between the two. The need to balance the currents to obtain precise half stepping is clearly a drawback to this scheme. Paradoxically, however, the properties of the machine with unequal phase currents can sometimes be turned to good effect, as we now see.

Step division – mini-stepping

There are some applications (e.g. in printing and photo-typesetting) where very fine resolution is called for, and a motor with a very small step angle – perhaps only a fraction of a degree – is required. We have already seen that the step angle can only be made small by increasing the number of rotor teeth and/or the number of phases, but in practice it is inconvenient to have more than four phases, and it is difficult to manufacture rotors with more than 50–100 teeth. This means it is rare for motors to have step angles below about 1°. When a smaller step angle is required a technique known as mini-stepping (or step division) is used.

Mini-stepping is a technique based on 2-phase-on opera-tion which provides for the sub-division of each full motor step into a number of 'substeps' of equal size. In contrast with half-stepping, where the two currents have to be kept equal, the currents are deliberately made unequal. By correctly choosing and controlling the relative amplitudes of the currents, the rotor equilibrium position can be made to lie anywhere between the step positions for each of the two separate phases.

Closed-loop current control is needed to prevent the current from changing as a result of temperature changes in the windings, or variations in the supply voltage, and if it is necessary to ensure that the holding torque stays constant for each ministep both currents must be changed in a quite complicated fashion. Despite the difficulties referred to above, mini-stepping is used extensively, especially in photographic and printing applications where a high resolution is needed. Schemes involving between 3 and 10 ministeps for a 1.8° step motor are numerous, and there are instances where up to 100 ministeps (20000 ministeps/rev) have been successfully achieved.

So far, we have concentrated on those aspects of behaviour which depend only on the motor itself, i.e. the static performance. The shape of the static torque curve, the holding torque, and the slope of the torque curve about the step position have all been shown to be important pointers to the way the motor can be expected to perform. All of these characteristics depend on the current(s) in the windings, however, and when the motor is running the instantaneous currents will depend on the type of drive circuit employed, as we will see next.

DRIVE CIRCUITS AND STEADY-STATE CHARACTERISTICS

Users often find difficulty in coming to terms with the fact that the running performance of a stepping motor depends so heavily on the type of drive circuit being used. It is therefore important to emphasize that in order to meet a specification, it will always be necessary to consider the motor and drive together as a package. We will therefore look at the reasons why the drive circuit is so influential, and at the most important dynamic characteristics of motor/drive systems.

Requirements of drive

The basic function of the complete drive is to convert the step command input signals into appropriate patterns of currents in the motor windings. This is achieved in two distinct stages, as shown in Figure 8.10, which relates to a 3-phase motor.

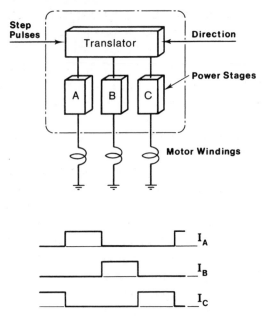

Figure 8.10   *General arrangement of drive system, and winding currents corresponding to an 'ideal' drive*

The 'translator' stage converts the incoming train of step command pulses into a sequence of on/off commands to each of the three power stages. In the 1 phase-on mode, for example, the first step command pulse will be routed to turn on phase A, the second will turn on phase B, and so on. In a very simple drive, the translator will probably provide for only one mode of operation (e.g. 1-phase-on), but most commercial drives provide the option of 1-phase-on. Single-chip ICs with these 3 operating modes, and with both 3-phase and 4-phase outputs are readily available.

The power stages (one per phase) supply the current to the windings. There is an enormous diversity of types in use, ranging from simple ones, with one switching transistor per phase, to elaborate chopper-type circuits with four transistors per phase, so it is helpful to list the functions required of the ideal power stage, before looking at how well the most important types of drive meet these ideal objectives.

The basic requirements are firstly that when the translator calls for a phase to be energized, the full (rated) current should be established immediately. Secondly, the current should be maintained constant (at its rated value) for the duration of the 'on' period. And finally, when the translator calls for the current to be turned off, it should be reduced to zero immediately.

The ideal current waveforms for continuous stepping with 1-phase-on operation are shown in the lower part of Figure 8.10. The currents have a square profile because this leads to the optimum value of running torque from the motor. Because of the inductance of the windings, no real drive will achieve the ideal current wave-forms, but many drives come close to the ideal, even at quite high stepping rates. Drives which produce such rectangular current waveforms are (not surprisingly) called constant-current drives. We now look at the running torque produced by a motor when operated from an ideal constant current drive. This will act as a yardstick for assessing the performance of other drives, all of which will be seen to have inferior performance.

Pull-out torque under constant-current conditions

If the phase currents are taken to be ideal, i.e. they are switched on and off instantaneously, and remain at their full rated value during each 'on' period, we can picture the axis of the magnetic field to be advancing around the machine in a series of steps, the rotor being urged to follow it by the reluctance torque. If we assume that the inertia is high enough for fluctuations in rotor velocity to be very small, the rotor will be rotating at a constant rate which corresponds exactly to the stepping rate.

Now if we consider a situation where the position of the rotor axis is, on average, lagging behind the advancing field axis, it should be clear that, on average, the rotor will experience a driving torque. The more it lags behind, the higher will be the average forward torque acting on it, but only up to a point. We already know that if the rotor axis is displaced too far from the field axis, the torque will begin to diminish, and eventually reverse, so we conclude that although more torque will be developed by increasing the rotor lag angle, there will be a limit to how far this can be taken.

Turning now to a quantitative examination of the torque on the rotor, we will make use of the static torque-displacement curves discussed earlier, and look at what happens when the load on the shaft is varied, the stepping rate being kept constant. Clockwise rotation will be studied, so the phases will be energized in the sequence ACB. The instantaneous torque on the rotor can be arrived at by recognizing (a) that the rotor speed is constant, and it covers one step angle (30°) between step command pulses, and (b) the rotor will be 'acted on' sequentially by each of the set of torque curves.

When the load torque is zero, the nett torque developed by the rotor must be zero (apart from a very small torque required to overcome friction). This condition is shown in Figure 8.11(a). The instantaneous torque is shown by the heavy line, and it is clear that each phase in turn exerts first a clockwise torque, then an anticlockwise torque while the rotor angle turns through 30°. The average torque is zero, the same as the load torque, because the average rotor lag angle is zero.

When the load torque on the shaft is increased, the immediate effect is to cause the rotor to fall back in relation to the field. This causes the clockwise torque to increase, and the anticlockwise torque to decrease. Equilibrium is reached when the lag angle has increased sufficiently for the average motor torque to equal the load torque. The torque developed at an intermediate load condition like this is shown in Figure 8.11(b). The highest torque that can possibly be developed is shown by the heavy line in Figure 8.11(c), and the average value of this torque is referred to as the pull-out torque. Since we have assumed an ideal constant-current drive, the pull-out torque will be independent of the stepping rate, and the pull-out torque-speed curve under ideal conditions is therefore as shown in Figure 8.12. The

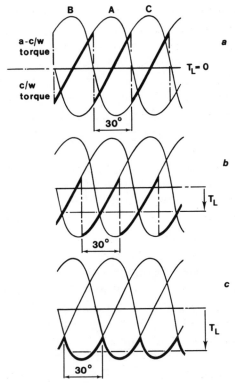

Figure 8.11   *Static torque curves indicating how the average steady-state torque is developed during constant-frequency operation*

shaded area represents the permissible operating region: at any particular speed (stepping rate) the load torque can have any value up to the pull-out torque, and the motor will continue to run at the same speed. But if the load torque exceeds the pull-out torque, the motor will suddenly stall.

As mentioned earlier, no real drive will be able to provide the ideal current waveforms, so we now turn to look briefly at the types of drives in common use, and at their torque-speed characteristics.

Drive circuits and pull-out torque-speed curves

There are three commonly-used types of drive. All use transistors which are operated as switches, i.e. they are either

Torque

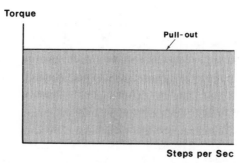

Pull-out

Steps per Sec

Figure 8.12 *Steady-state operating region with ideal constant-current drive.* *(In such idealized circumstances there would be no limit to the stepping rate, but as Figure 8.14 shows any real drive circuit imposes an upper limit.)*

turned fully on, or they are cut-off. A brief description of each is given below, and the pros and cons of each type are indicated. In order to simplify the discussion, we will consider one phase of a 3-phase VR motor and assume that it can be represented by a simple R-L circuit, though in practice the inductance will vary with rotor position, giving rise to motional e.m.f. in the windings.

Constant voltage drive

This is the simplest possible drive, and is shown in Figure 8.13(a). The d.c. voltage V is chosen so that when the transistor is on, the steady current (V/R) is the rated current, and the current rises exponentially with a time-constant of L/R. The freewheel diode (see Chapter 2) only conducts for a relatively short period at the end of each 'on' period, after the transistor is turned off. Since the winding is inductive, the current through it cannot be reduced to zero instantaneously, so when the transistor turns off, the current is diverted into the closed path formed by the winding and the diode, and it then decays exponentially to zero, with time-constant L/R.

At low stepping rates, the drive provides a reasonably good (i.e. rectangular) current waveform, as shown in Figure 8.13(a). But at higher frequencies, where the 'on' period is

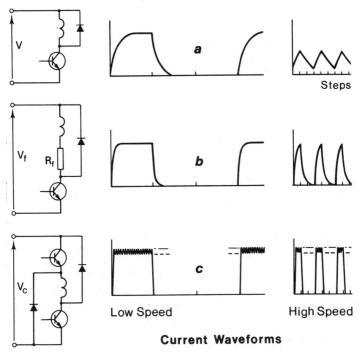

Figure 8.13   *Drive circuits and corresponding motor current waveforms*

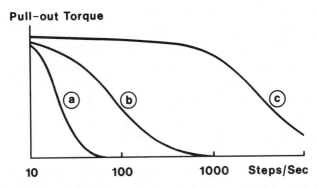

Figure 8.14   *Typical pull-out torque-speed curves for a given motor with different types of drive circuit (a) constant voltage drive; (b) current-forced drive; (c) chopper drive*

short compared with the winding time-constant, the current waveform degenerates, and is nowhere near to the ideal shape. The pull-out torque therefore falls off very rapidly, as shown in Figure 8.14(a), which means that the motor is restricted to low-speed operation.

Current-forced drive

In order to increase the rate of rise of current at switch-on, a higher supply voltage is needed. However, to prevent the current from exceeding the rated value, an additional 'forcing' resistor has to be added in series with the winding. The value of this resistance (R_f) must be chosen so that $V_f/(R + R_f) = I$, where I is the rated current. This is shown in Figure 8.13(b), together with the current waveforms. Because the rates of rise and fall of current are higher, the current waveforms are much improved, especially at high stepping rates, as can be seen. The pull-out torque is therefore maintained to a higher stepping rate, as shown in Figure 8.14(b). Values of R_f from two to ten times the motor resistance (R) are common. Broadly speaking, if $R_f = 10R$, a given pull-out torque will be available at ten times the stepping rate, compared with an unforced constant voltage drive.

Some manufacturers call this type of drive an 'R/L' drive, while others call it an 'L/R' drive, or even simply a 'Resistor Drive'. Often, sets of torque speed curves in catalogues are labelled with values R/L (or L/R) = 5, 10 etc. This means that the curves apply to drives where the forcing resistor is five (or ten) times the winding resistance, the implication being that the drive voltage has also been adjusted to keep the static current at its rated value. Obviously, it follows that the higher R_f is made, the higher the power rating of the supply; and it is the higher power rating which is the principal reason for the improved torque-speed performance.

The major disadvantage of this drive is its inefficiency, and the consequent need for a high power-supply rating. Large amounts of heat are dissipated in the forcing resistors, especially when the motor is at rest and the phase-current is continuous, and disposing of the heat can lead to awkward problems in the siting of the forcing resistors.

It was mentioned earlier that the influence of the motional e.m.f. in the winding would be ignored. In practice, however, the motional e.m.f. always has a pronounced influence on the current, especially at high stepping rates, so it must be borne in mind that the waveforms shown in Figure 8.13(a & b) are only approximate. Not surprisingly, it turns out that the motional e.m.f. tends to make the current waveforms worse (and the torque less) than the discussion above suggests. Ideally therefore, we need a drive which will keep the current constant throughout the on period, regardless of the motional e.m.f. The closed-loop chopper-type drive (below) provides the closest approximation to this, and also avoids the waste of power which is a feature of R/L drives.

Chopper drives

The basic circuit for one phase is shown in Figure 8.13(c) together with the current wave-forms. A high-voltage power supply is used, with two switching transistors per phase. The lower transistor is turned on for the whole period during which current is required. The upper transistor turns on whenever the actual current falls below the lower threshold (shown dotted in Figure 8.13(c)), and it turns off when the current exceeds the upper threshold. The chopping action leads to a current waveform which is a good approximation to the ideal. Because the current-control system is a closed-loop one, distortion of the current waveform by the motional e.m.f. is minimized, and this means that the ideal (constant-current) torque-speed curve is closely followed up to high stepping rates. Eventually, however, the 'on' period reduces to the point where it is less than the current rise time, and the full current is never reached. Chopping action then ceases, the drive reverts essentially to a constant-voltage one, and the torque falls rapidly as the stepping rate is raised even higher, as in Figure 8.14(c).

There is no doubt of the overall superiority of the chopper-type drive, and it will doubtless eventually become the standard drive. Single-chip chopper modules can be bought for small (say 1–2 A) motors, and complete plug-in chopper cards, rated up to 10A or more are available.

Resonances and instability

In practice, measured torque-speed curves frequently display severe dips at or around certain stepping rates. Manufacturers are not keen to stress this feature, and sometimes omit the dips from their curves, so it is doubly important for the user to be on the lookout for them. A typical measured characteristic is shown in Figure 8.15(a).

The magnitude and location of the torque dips depend in a complex way on the characteristics of the motor, the drive, the operating mode and the load. We will not go into detail here, apart from mentioning the underlying causes and remedies.

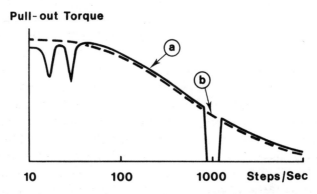

Figure 8.15   *Pull-out torque-speed curves for a hybrid stepping motor showing (curve a) low-speed resonance dips, mid-frequency instability at around 1000 steps/sec, and improvement brought about by adding an inertia damper (curve b)*

There are two distinct mechanisms which cause the dips. The first is a straightforward 'resonance-type' problem which manifests itself at low stepping rates, and originates from the oscillatory nature of the single-step response. Whenever the stepping rate coincides with the natural frequency of rotor oscillations, the oscillations can be enhanced, and this in turn makes it more likely that the rotor will fail to keep in step with the advancing field.

The second phenomenon occurs because at certain stepping rates it is possible for the complete motor/drive system to exhibit positive feedback, and become unstable. This instability usually occurs at relatively high stepping rates, well above the resonance regions discussed above. The resulting dips in the torque-speed curve are extremely sensitive to the degree of viscous damping present (mainly in the bearings), and it is not uncommon to find that a severe dip which is apparent on a warm day will disappear on a cold day.

The dips are most pronounced during steady-state operation, and it may be that their presence is not serious as long as continuous operation at the relevant speeds is not required. In this case, it is often possible to accelerate through the dips without adverse effect. Various special drive techniques exist for eliminating resonances by smoothing out the step-wise nature of the stator field, or by modulating the supply frequency to damp out the instability, but the simplest remedy in open-loop operation is to fit a damper to the shaft. Dampers of the Lanchester type (or of the viscously-coupled inertia (VCID) type) are used. These have the advantage that they are cheap and robust, and do not exert a drag torque when the rotor speed is constant because their damping torque comes into play only when the rotor is changing speed. The drawback is that they increase the effective inertia of the system, and thus reduce the maximum acceleration. By selecting an appropriate damper, the torque dips can be eliminated, as shown in Figure 8.15(b). Dampers are also often essential to damp the single-step response, particularly with VR type motors, many of which have a highly oscillatory step response.

TRANSIENT PERFORMANCE

Step response

It was pointed out earlier that the single-step response is similar to that of a damped second-order system. We can easily estimate the natural frequency (ω_n) from the equation

$$\omega_n^2 = \frac{\text{slope of torque-angle curve}}{\text{total inertia}}.$$

Knowing ω_n, we can judge what the first part of the response will look like, by assuming the system is undamped. To refine the estimate, and to obtain the settling time, however, we need to estimate the damping ratio, which is much more difficult to determine, as it depends on the type of drive circuit and mode of operation as well as on the mechanical friction. In VR motors the damping ratio can be as low as 0.1, but in hybrid types it is typically 0.3–0.4. These values are too low for many applications where rapid settling is called for.

Two remedies are available, the simplest being to fit a mechanical damper of the type mentioned above. Alternatively, a special sequence of timed command pulses can be used to brake the rotor so that it reaches its new step position with zero velocity and does not overshoot. This procedure is variously referred to as 'electronic damping', 'electronic braking' or 'back phasing'. It involves re-energizing the previous phase for a precise period before the rotor has reached the next step position, in order to exert just the right degree of braking. It can only be used successfully when the load torque and inertia are predictable and not subject to change. Because it is an open-loop scheme it is extremely sensitive to apparently minor changes such as day-to-day variation in friction, which can make it unworkable in many instances.

Starting from rest

The rate at which the motor can be started from rest without losing steps is known as the 'starting' or 'pull-in' rate. The starting rate for a given motor depends on the type of drive, and the parameters of the load. This is entirely as expected since the starting rate is a measure of the motor's ability to accelerate its rotor and load and pull into synchronism with the field. The starting rate thus reduces if either the load torque, or the load inertia are increased. Typical pull-in torque-speed curves, for various inertias, are shown in Figure 8.16. The pull-out torque speed curve is also shown, and it can be seen that for a given load torque, the maximum steady (slewing) speed at which the motor can run is much higher than the corresponding starting rate. (Note that only one pull-out torque is usually shown, and is taken to apply for all inertia values. This is because the inertia is not significant when the speed is constant.)

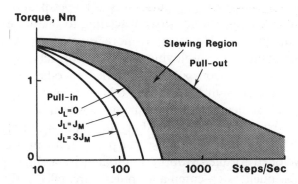

Figure 8.16 *Typical pull-in and pull-out curves showing effect of load inertia on the pull-in torque. (Load inertia = J_L; motor inertia = J_M)*

It will normally be necessary to consult the manufacturer's data to obtain the pull-in rate, which will apply only to a particular drive. However, a rough assessment is easily made: we simply assume that the motor is producing its pull-out torque, and calculate the acceleration that this would produce, making due allowance for the load torque and inertia. If, with the acceleration as calculated, the motor is able to reach the steady speed in one step or less, it will be able to pull in; if not, a lower pull-in rate is indicated.

Optimum acceleration and closed-loop control

There are some applications where the maximum possible accelerations and decelerations are demanded, in order to minimize point-to-point times. This calls for the whole train of step command pulses to be delivered at progressively shorter intervals during acceleration, the intervals being chosen in a precise way. Each phase must only be on when it can produce positive torque, but since the torque depends on the rotor position, the optimum switching times have to be calculated from a full dynamic analysis. This can usually be accomplished by making use of the static torque-angle curves, provided appropriate allowance is made for the rise and fall times of the stator currents.

The train of accelerating pulses has to be pre-programmed into the controller, for subsequent feeding to the drive in an open-loop fashion. However, a more satisfactory arrangement is obtained by employing a closed-loop scheme, using position feedback from a shaft-mounted encoder or by deducing position from the motor's own windings. The feedback signals indicate the instantaneous position of the rotor, and can be used to ensure that the phase-windings are switched at precisely the right rotor position for maximizing the developed torque. Motion is initiated by a single pulse, and subsequent step-command pulses are effectively self-generated by the encoder. The motor continues to accelerate until its torque equals the load torque, and then runs at this (maximum) speed until the deceleration sequence is initiated. During all this time, the step counter continues to record the number of steps taken.

Closed-loop operation ensures that the optimum acceleration is achieved, but at the expense of more complex control circuitry, and the need to fit a shaft encoder. Relatively cheap encoders are however now available for direct fitting to some ranges of motors, and single chip microprocessor-based controllers are available which provide all the necessary facilities for closed-loop control.

Encoders are also used in open-loop schemes when an absolute check on the number of steps taken is required. Used in this way the encoder simply provides a tally of the total steps taken, and normally need play no part in the generation of the step pulses. At some stage, however, the actual number of steps taken will be compared with the number of step command pulses issued by the controller. When a disparity is detected, indicating a loss (or gain of steps), the appropriate additional forward or backward pulses can be added.

9

SYNCHRONOUS, SWITCHED RELUCTANCE AND BRUSHLESS D.C. DRIVES

INTRODUCTION

In this chapter the common feature which links the motors is that they are all, in one way or another, a.c. motors. In terms of market share none of them compares with the induction motor, but some of the drive schemes which we look at show great promise, and are likely to become more important in the future. The idea of using synchronous or reluctance machines in controlled drives is not new, but until power electronic converters became available it was not practicable to implement such schemes. Nowadays the trend is increasingly towards tightly-integrated motor/converter packages such as the brushless d.c. drive, where the motor can only be used with its own converter. To begin with, however, we will look at synchronous motors which can be used either directly on the 50 Hz or 60 Hz mains, or supplied from a converter.

SYNCHRONOUS MOTORS

In Chapter 5 we saw that the 3-phase stator windings of an induction motor produce a sinusoidal rotating magnetic field in the air-gap. The speed of rotation of the field (the synchronous speed) was shown to be directly proportional to the supply frequency, and inversely proportional to the pole-number of the winding. We also saw that in the induction motor the rotor is dragged along by the field, but that the higher the load on the shaft, the more the rotor has to slip with respect to the field in order to induce the rotor currents required to produce the torque. Thus although at no-load the speed of the rotor can be close to the synchronous speed, it must always be less; and as the load increases, the speed has to fall.

In the synchronous motor, the stator windings are exactly the same as in the induction motor, so when connected to the 3-phase supply, a rotating magnetic field is produced. But instead of having a cylindrical rotor with a cage winding, the synchronous motor has a rotor with either a d.c. excited winding, or permanent magnets, designed to cause the rotor to 'lock-on' or 'synchronize with' the rotating magnetic field produced by the stator. Once the rotor is synchronized, it will run at exactly the same speed as the rotating field despite load variation, so under mains-frequency operation the speed will remain constant as long as the mains frequency is stable.

We discussed a very similar process in connection with the stepping motor (Chapter 8), but there the field tended to proceed in a step-wise fashion, rather than smoothly. As with the stepper, we find that there is a limit to the maximum (pull-out) torque which can be developed before the rotor is forced out of synchronism with the rotating field. Pull-out torque will typically be 1.5 times the continuous rated torque, but for all torques below pull-out the speed will be absolutely constant. The torque-speed curve is therefore simply a vertical line at the synchronous speed, as shown in Figure 9.1 We can see from Figure 9.1 that the vertical line extends into quadrant 2, which indicates that if we try to force the speed above the synchronous speed the machine will act as a generator.

The mains-fed synchronous motor is clearly ideal where a constant speed is essential, and also where several motors must run at precisely the same speed. Examples where

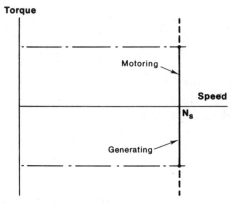

Figure 9.1 *Steady-state torque-speed curve for a synchronous motor supplied at constant frequency*

3-phase motors are used include artificial fibre spinning lines, and film and tape transports. Small single-phase reluctance versions are used in clocks and timers for washing machines, heating systems etc. They are also used where precise integral speed ratios are to be maintained: for example a 3:1 speed ratio can be guaranteed by using a 2-pole and a 6-pole motor, fed from the same supply.

We will now look briefly at the various types of synchronous motor, mentioning the advantages and disadvantages of each. One obvious question which must be addressed is how such motors run up to synchronous speed. This often turns out to be a problem, especially for large motors driving high-inertia loads.

Excited rotor motors

The rotor carries a 'field' winding which is supplied with direct current via a pair of sliprings on the shaft, and is designed to produce an air-gap field of the same pole-number and spatial distribution (usually sinusoidal) as that produced by the stator winding. The rotor may be cylindrical, with the field winding distributed in slots, or it may have projecting ('salient') poles around which the winding is concentrated (Figure 9.2).

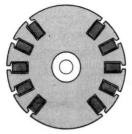
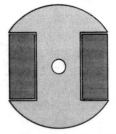

Figure 9.2 *Rotors for synchronous motors. 2-pole cylindrical (left) with field coils distributed in slots, and 2-pole salient pole (right) with concentrated field winding*

The simplest way to visualize the mechanism of torque production is to focus on a static picture, and consider the alignment force between the stator and rotor field patterns. When the two are aligned with N facing S, the torque is zero and the system is in stable equilibrium, with any displacement to right or left causing a restoring torque to come into play. If the fields are distributed sinusoidally in space, the restoring torque will reach a maximum or 'pull-out' value when the poles are misaligned by half a pole-pitch, or 90°. Beyond 90° the torque reduces with angle, giving an unstable region, zero torque being reached again when N is opposite N.

When the motor is running matters are rather more complex than the simplistic view taken above might suggest, because the strength of the stator field depends on the stator current, which in turn varies with both the rotor excitation and the load. But the notion of the rotor running so that its field lags behind the resultant field by an angle which increases with load is adequate for our purposes. Each time the load increases, the rotor slows momentarily before settling at the original speed but with an increased 'load-angle'. We can actually see this happen if we illuminate the shaft of the motor with a mains-frequency stroboscope: a reference mark on the shaft is seen to drop back by a few degrees each time the load is increased.

Excited rotor motors are used in sizes ranging from a few kW up to several MW. The large ones are effectively alternators (as used for power generation) but used as motors. Wound rotor induction motors (see Chapter 6) can also be made to operate synchronously by supplying the rotor with d.c. through the sliprings. In all cases the rotor 'excitation' power is relatively small, since all the mechanical output power is supplied from the stator side.

Starting

It should be clear from the discussion of how torque is produced that unless the rotor is running at the same speed as the rotating field, no steady torque can be produced. If the rotor is running at a different speed, the two fields will be sliding past each other, giving rise to a pulsating torque with an average value of zero. Hence a simple synchronous machine is not self-starting, and some alternative method of producing a run-up torque is required.

Most synchronous motors are therefore equipped with some form of rotor cage, similar to that of an induction motor, in addition to the main field winding. When the motor is switched onto the mains supply, it operates as an induction motor during the run-up phase, until the speed is just below synchronous. The excitation is then switched on so that the rotor is able to make the final acceleration and 'pull-in' to synchronism with the rotating field. Because the cage is only required during starting, it can be short-time rated, and therefore comparatively small. Once the rotor is synchronized, and the load is steady, no currents are induced in the cage, because the slip is zero. The cage does however come into play when the load changes, when it provides an effective method for damping out the oscillations of the rotor as it settles at its new steady-state load angle.

Large motors will tend to draw a very heavy current during run-up, so some form of reduced voltage starter is often required (see Chapter 6). Sometimes, a separate small induction motor is used simply to run-up the main motor prior to synchronization, but this is only feasible where the load is not applied until after the main motor has been synchronized.

No special starter is required for the wound rotor induction motor of course, which runs up in the usual way (see Chapter 6) before the d.c. excitation is applied. Motors operated like this are sometimes known as 'inductosyns'.

Power factor control

If the rotor excitation is varied when the motor is running on load, the load angle and stator current change, but the speed, load (and hence input power) remain the same. The power factor of the motor can therefore be adjusted by means of the excitation. Unity power factor operation is possible over a wide range of loads, and by using a high rotor excitation the motor can operate at leading power factor. This is particularly valuable with a large synchronous motor, which can be used to compensate for the lagging power factor of other induction motors on the same site. At the other extreme, there is a limit to how low the excitation can be set, because if the rotor field is too weak it will not be capable of developing the required load torque, even at the optimum load angle.

Permanent-magnet synchronous motors

Permanent magnets can be used on the rotor instead of a wound field, as shown in Figure 9.3. A wide variety of configurations is possible, and for those shown in Figure 9.3 the

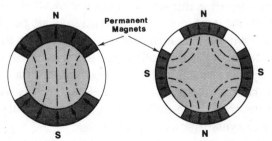

Figure 9.3   *Permanent magnet synchronous motor rotors 2-pole (left); 4-pole (right)*

direction in which the magnets have been magnetized is shown by the arrows. Motors of this sort have outputs ranging from about 100 W up to perhaps 100 kW.

For starting from a fixed-frequency supply a rotor cage is required, as discussed above. The advantages of the permanent-magnet type are that no supply is needed for the rotor and the rotor construction can be robust and reliable. The disadvantage is that the excitation is fixed, so the designer must either choose the shape and disposition of the magnets to match the requirements of one specific load, or seek a general-purpose compromise.

Early permanent-magnet motors suffered from the tendency for the magnets to be demagnetized by the high stator currents during starting, and from a restricted maximum allowable temperature. Much improved versions using high coercivity rare-earth magnets were developed during the 1970s to overcome these problems. They are usually referred to as 'Line-Start' motors, to indicate that they are designed for direct-on-line starting. The steady-state efficiency and power factor at full load are in most cases better than the equivalent induction motor, and they can pull-in to synchronism with inertia loads of many times rotor inertia.

Hysteresis motors

Whereas most motors can be readily identified by inspection when they are dismantled, the hysteresis motor is likely to baffle anyone who has not come across it before. The rotor consists simply of a thin-walled cylinder of what looks like steel, while the stator has a conventional single-phase or three-phase winding. Evidence of very weak magnetism may just be detectable on the rotor, but there is no hint of any hidden magnets as such, and certainly no sign of a cage. Yet the motor runs up to speed very sweetly and settles at exactly synchronous speed with no sign of a sudden transition from induction to synchronous operation.

These motors (the operation of which is quite complex) rely mainly on the special properties of the rotor sleeve, which is made from a hard steel which exhibits pronounced magnetic hysteresis. Normally in machines we aim to minimize hysteresis in the magnetic materials, but in these motors the effect (which arises from the fact that the magnetic flux density B depends on the previous 'history' of the MMF) is deliberately accentuated to produce torque. There is actually also some induction motor action during the run-up phase, and the nett result is that the torque remains roughly constant at all speeds.

Small hysteresis motors are used extensively in tape recorders, office equipment, fans etc. The near constant torque during run-up and the very modest starting current (of perhaps 1.5 times rated current) means that they are also suited to high inertia loads such as gyro compasses and small centrifuges.

Reluctance motors

The reluctance motor is arguably the simplest synchronous motor of all, the rotor consisting simply of a set of iron laminations shaped so that it tends to align itself with the field produced by the stator. This 'reluctance torque' action was discussed when we looked at the variable reluctance stepping motor (in Chapter 8).

Here we are concerned with mains-frequency reluctance motors which differ from steppers in that they only have saliency on the rotor, the stator being identical with that of a 3-phase induction motor. In fact, since induction motor action is required in order to get the rotor up to synchronous speed, a practical reluctance-type rotor does have more than just a shaped rotor, and in practice it resembles a cage induction motor. However parts of the periphery are cut away in order to force the flux from the stator to enter the rotor in the remaining regions where the air-gap is small, as shown in Figure 9.4(a). Alternatively, the 'preferred flux paths' can be imposed by removing iron inside the rotor so that the flux is guided along the desired path, as shown in Figure 9.4(b).

The rotor will tend to align itself with the field, and hence

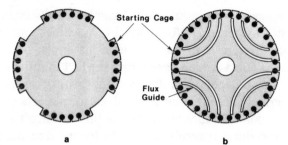

Figure 9.4 *Reluctance motor rotors (4-pole). (a) salient type and (b) flux-guided type*

is able to remain synchronized with the travelling field set up by the three-phase winding on the stator in much the same way as a permanent-magnet rotor. Early reluctance motors were invariably one or two frame sizes bigger than an induction motor for a given power and speed, and had low power-factor and poor pull-in performance. As a result they fell from favour except for some special applications such as textile machinery, where cheap constant speed motors were required. Understanding of reluctance motors is now much more advanced, and they can compete on almost equal terms with the induction motor as regards power-output, power factor and efficiency. They are nevertheless relatively expensive because they are not produced in large numbers.

CONTROLLED-SPEED SYNCHRONOUS MOTOR DRIVES

As soon as variable-frequency inverters became a practicable proposition it was a natural step to seek to use them to supply synchronous motors, thereby freeing them from the fixed-speed constraint imposed by mains-frequency operation and opening up the possibility of a simple open-loop controlled speed drive. The obvious advantage over the inverter-fed induction motor is that the speed of the synchronous motor is exactly determined by the frequency whereas the induction motor always has to run with a finite slip. A precision frequency source (oscillator) controlling the inverter switching is all that is necessary to give accurate speed control with a synchronous motor, while speed feedback is essential to achieve accuracy with an induction motor.

In practice open-loop operation of inverter-fed synchronous motors is not as widespread as might be expected, though it is commonly used in multi-motor drives (see below). Closed-loop or self-synchronous operation is however rapidly gaining momentum, and is already well established at the high- and low-power ends of the scale. At one extreme, large excited-rotor synchronous motors are used in place of d.c. drives, particularly where high speeds are required or when the motor must operate in a hazardous atmosphere (e.g. in a large gas compressor). At the other end of the scale, small permanent-magnet synchronous motors are used in brushless d.c. drives.

Open-loop inverter-fed synchronous motor drives

This simple method is attractive in multi-motor installations where all the motors must run at exactly the same speed. Individually the motors (permanent-magnet or reluctance) are more expensive than the equivalent mass-produced induction motor, but this is offset by the fact that speed feedback is not required, and the motors can all be supplied from a single inverter, as shown in Figure 9.5.

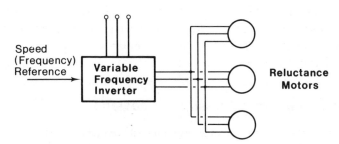

Figure 9.5 *Open-loop operation of a group of several reluctance motors supplied from a single variable-frequency inverter*

The inverter voltage/frequency ratio will usually be kept constant (see Chapter 7) to ensure that the motors operate at full flux at all speeds, and therefore have a 'constant-torque' capability. If prolonged low speed operation is called for, improved cooling of the motors may be necessary. Speed is precisely determined by the inverter frequency, but speed changes (including run-up from rest) must be made slowly, under ramp control, to avoid the possibility of exceeding the pull-out load angle, which would result in stalling.

A problem which can sometimes occur with this sort of open-loop operation is that the speed of the motor exhibits apparently spontaneous oscillation or 'hunting'. The supply frequency may be absolutely constant but the rotor speed is seen to fluctuate about its expected (synchronous) value, sometimes with an appreciable amplitude, and usually at a low frequency of perhaps 1 Hz. The origin of this unstable behaviour lies in the fact that the motor and load constitutes at least a fourth-order system, and can therefore become very poorly damped or even unstable for certain combinations of the system parameters. Factors which influence stability are terminal voltage, supply frequency, motor time-constants and load inertia and damping. Unstable behaviour in the strict sense of the term (i.e. where the oscillations build up without limit) is rare, but bounded instability is not uncommon, especially at speeds well below the base level (50 Hz or 60 Hz), and under light-load conditions. It is very difficult to predict exactly when unstable behaviour might be encountered, and provision must be made to combat it. Some inverters therefore include circuitry which detects any tendency for the currents to fluctuate (indicating hunting) and to modulate the voltage and/or frequency to suppress the unwanted oscillations.

Self-synchronous (closed-loop) operation

In the open-loop scheme outlined above, the frequency of the supply to the motor is under the independent control of the oscillator driving the switching devices in the inverter.

The inverter has no way of knowing whether the rotor is correctly locked-on to the rotating field produced by the stator, and if the pull-out torque is exceeded, the motor will simply stall.

In the self-synchronous mode, however, the inverter output frequency is determined by the speed of the rotor. More precisely, the instants at which the switching devices operate to turn the stator windings on and off are determined by rotor-position-dependent signals obtained from an encoder mounted on the rotor shaft. In this way, the stator currents are always switched on at the right time to produce the desired torque on the rotor, because the inverter effectively knows where the rotor is at every instant of time. The use of rotor postion feedback signals to control the inverter accounts for the description 'closed-loop' used above. If the rotor slows down (as a result of an increase in load, for example), the stator supply frequency automatically reduces so that the rotor remains synchronized with the field, and the motor therefore cannot 'pull-out' in the way it does under open-loop operation.

An analogy with the internal combustion engine may help to clarify the difference between closed-loop and open-loop operation. An engine invariably operates as a closed-loop system in the sense that the opening and closing of the inlet and exhaust valves is automatically synchronized with the position of the pistons by means of the camshaft and timing chain or belt. The self-synchronous machine is much the same in that the switching devices in the inverter turn the current on and off according to the position of the rotor. By contrast, open-loop operation of the engine would imply that we had removed the timing chain and chosen to operate the valves by driving the camshaft independently, in which case it should be clear that the engine would only be capable of producing power at the one speed at which the up and down motion of the pistons corresponded exactly with the opening and closing of the valves.

It turns out that the overall operating characteristics of a self-synchronous a.c. motor are very similar to those of a conventional d.c. motor. This is really not surprising when we recall that in a d.c. motor, the mechanical commutator reverses the direction of the current in each (rotating) armature coil at the appropriate point such that, regardless of speed, the current under each (stationary) field pole is always in the right direction to produce the desired torque. In the self-synchronous motor the roles of stator and rotor are reversed compared with the d.c. motor. The field is rotating and the 'armature' winding (consisting of three discrete groups of coils or phases) is stationary. The timing and direction of the current in each phase is governed by the inverter switching, which in turn is determined by the rotor position sensor. Hence regardless of speed, the torque is always in the right direction.

The combination of the rotor position sensor and inverter performs effectively the same function as the commutator in a conventional d.c. motor. There are of course usually only thee windings to be switched by the inverter, as compared with many more coils and commutator segments to be switched by the brushes in the d.c. motor, but otherwise the comparison is valid. Not surprisingly the combination of position sensor and inverter is sometimes referred to as an 'electronic commutator', while the overall similarity of behaviour gives rise to the rather clumsy term 'electronically commutated motor' (ECM) or the even worse 'commutator-less d.c. motor' (CLDCM) to describe self-synchronous machines.

Operating characteristics and control

If the d.c. input to the inverter is kept constant and the motor starts from rest, the motor current will be large at first, but will decrease with speed until the motional e.m.f. generated inside the motor is almost equal to the applied voltage. When the load on the shaft is increased, the speed begins to fall, the motional e.m.f reduces and the current increases until a new equilibrium is reached where the extra motor torque is equal to the load torque. This behaviour parallels that of the conventional d.c. motor, where the no-load speed depends on the applied armature voltage. The speed of the self-synchronous motor can therefore be controlled by controlling the d.c. link voltage to the inverter. The d.c. link will usually be provided by a controlled rectifier, so the motor speed can be controlled by varying the input converter firing angle, as shown in Figure 9.6.

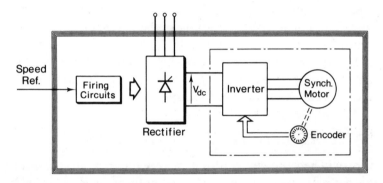

Figure 9.6   *Self-synchronous motor/inverter system. In large sizes this arrangement is sometimes referred to as a 'synchdrive'; in smaller sizes it would be known as a brushless d.c. motor drive*

The overall similarity with the d.c. drive (Chapter 4) is deliberately emphasized in Figure 9.6. The dotted line enclosing the a.c. motor together with its rotor position detector and inverter is in effect the replacement for the conventional d.c. motor. We note that a tachogenerator is not necessary for closed-loop speed control because the speed feedback signal can be derived from the frequency of the rotor position signal. And, as with the d.c. drive, current control, as distinct from voltage control can be used where the output torque rather than speed is to be controlled. Full four-quadrant operation is possible, as long as the inverter is supplied from a fully-controlled converter.

In simple cost terms the self-synchronous system looks attractive when the combined cost of the inverter and synchronous motor is lower than the equivalent d.c. motor.

When such schemes were first introduced (in the 1970s) they were only cost-effective in very large sizes (say over 1 MW), but the break-even point for wound-field motors is falling steadily and drives with ratings in the tens of kW are now becoming popular. They can utilize converter-grade (relatively cheap) thyristors in the inverter bridge because the thyristors will commutate naturally with the aid of the motor's generated e.m.f. At very low speeds however, the generated e.m.f. is insufficient, so the motor is started under open-loop current-fed operation, in the manner of a stepping motor.

As transistor inverter costs have fallen, lower power drives using permanent-magnet motors have become attractive, especially where very high speeds are required.

SWITCHED RELUCTANCE MOTOR DRIVES

The switched reluctance (SR) drive is the newest arrival on the drives scene, and can offer advantages in terms of efficiency, power per unit weight and volume, robustness and operational flexibility. The drawbacks (which it shares with the other relatively new self-synchronous systems) are that it is relatively unproven, can be noisy, and is inherently not well-suited to smooth torque production. Despite being still in its infancy, SR technology has been successfully applied to a wide range of applications including general purpose industrial drives, traction, domestic appliances and office and business equipment.

Principle of operation

The switched reluctance motor differs from the conventional 3-phase induction motor in that both the rotor and the stator have salient poles. This doubly-salient arrangement (as shown in Figure 9.7) proves to be very effective as far as electromagnetic energy conversion is concerned.

The stator carries coils on each pole, the coils on opposite poles being connected in series. The rotor, which is made

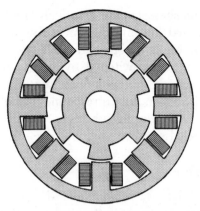

Figure 9.7 *Typical switched reluctance (SR) motor. Each of the eight stator poles carries a concentrated winding, while the six-pole rotor has no windings or magnets*

from laminations in the usual way, has no windings or magnets and is therefore cheap to manufacture and extremely robust. The particular example shown in Figure 9.7 has eight stator poles and six rotor poles, and represents a widely used arrangement, but other pole combinations are used to suit different applications. In Figure 9.7 the eight coils are grouped to form four phases, which are independently energized from a four-phase converter.

The motor rotates by energizing the phases sequentially in the sequence 1,2,3,4 for anticlockwise rotation or 1,4,3,2 for clockwise rotation, the 'nearest' pair of rotor poles being pulled into alignment with the appropriate stator poles by reluctance torque action. At low and medium speeds, a chopping action is used to maintain a rectangular current waveform, while at higher speeds the full voltage is applied throughout the 'on' period of each phase.

Any reader who is familiar with stepping motors (Chapter 8) will correctly identify the SR motor as a variable reluctance stepping motor. There are of course detailed design differences which reflect the different objectives (continuous rotation for the SR, stepwise progression for the stepper), but otherwise the mechanisms are identical. However whilst the stepper is designed first and foremost for open-loop operation, the SR motor is designed for self-synchronous operation, the phases being switched by signals derived from a shaft-mounted rotor position detector. This causes the behaviour of the SR motor to resemble that of a d.c. motor, in the same way as we saw above in the case of the rotor position controlled synchronous motor.

Power converter

An important difference between the SR motor and all other self-synchronous motors is that its full torque capability can be achieved without having to provide for both positive and negative currents in the phases. This welcome simplification arises because the direction of torque in a reluctance motor is independent of the direction of the current, and it means that the power converter can have fewer main switching devices (4 in the example above) than the six required for 3-phase bipolar inverters. A further major benefit comes from the fact that because each of the main devices is permanently connected in series with one of the motor windings, there is no possibility of the 'shoot-through' fault (see Chapter 2) which is a major headache in the conventional inverter.

Characteristics and control

In common with other self-synchronous drives, a wide range of operating characteristics is available. If the input converter is fully controlled, continuous regeneration and full 4-quadrant operation is possible, and the usual constant torque, constant power and series type characteristic is regarded as standard. Low speed torque tends to be uneven unless special measures are taken to profile the current pulses, but continuous low speed operation is better than for most competing systems because of the high efficiency. Some of the early SR motors were deemed to be noisy, but improved mechanical design suggests that the SR will not be significantly worse than the competition in this sensitive area.

BRUSHLESS D.C. MOTORS

Much of the impetus for the recent development of brushless d.c. motors came from the computer-peripheral and aerospace industries, where high performance coupled with reliability and low maintenance are essential. Very large numbers of brushless d.c. motors are now used, particularly in sizes up to a few hundred watts. The small versions (less than 100 W) are increasingly made with all the control and power electronic circuits integrated at one end of the motor, so that they can be directly retro-fitted as a replacement for a conventional d.c. motor. Because all the heat-dissipating circuits are on the stator, cooling is much better than in a conventional motor, so higher specific outputs can be achieved. The rotor inertia can also be less than that of a conventional armature, which means that the torque/inertia

Plate 9.1   *Brushless d.c. motors with 4-pole rare earth permanent-magnet rotors (see Figure 9.3). These motors are intended for servo-type applications, and have integral Hall effect rotor position sensors (Photograph by courtesy of Evershed and Vignoles Ltd)*

ratio is better, giving a faster acceleration. Higher speeds are practicable because there is no mechanical commutator.

In principle, there is no difference between a brushless d.c. motor and the self-synchronous permanent-magnet motor discussed earlier in this chapter. The reader may therefore be puzzled as to why some motors are described as brushless d.c. while others are not. In fact, there is no logical reason at all, nor indeed is there any universal definition or agreed terminology.

Broadly speaking, however, the accepted practice is to restrict the term 'brushless d.c. motor' to a particular type of self-synchronous permanent-magnet motor in which the rotor magnets and stator windings are arranged to produce a trapezoidal air-gap flux. Such motors are fed from inverters which produce rectangular current waveforms, the switch-on being initiated by digital signals from a relatively simple rotor position sensor. This combination permits the motor to develop a more or less smooth torque, regardless of speed, but does not require an elaborate position sensor. (In contrast, many self-synchronous machines have sinusoidal air-gap fields, and therefore require more sophisticated position sensing if they are to develop smooth torque.)

Characteristics and control

The brushless d.c. motor is essentially an inside-out electronically commutated d.c. motor, and can therefore be controlled in the same way as a conventional d.c. motor (see Chapter 4). Many brushless motors are used in demanding servo-type applications, where they need to be integrated with digitally controlled systems. For this sort of application, complete digital control systems which provide for torque, speed and position control are available.

10

MOTOR/DRIVE SELECTION

INTRODUCTION

Experience suggests that three main areas of difficulty are often encountered in the selection process. Firstly, as we have discovered in the preceding chapters, there is a good deal of overlap between the major types of motor and drive. This makes it impossible to lay down a set of hard and fast rules to guide the user straight to the best solution for a particular application. Secondly, users tend to underestimate the importance of starting with a comprehensive specification of what they really want, and they seldom realize how much weight attaches to such things as the torque-speed curve and inertia of the load. And thirdly they may be unaware of the existence of standards and legislation, and hence can be baffled by questions from any potential supplier.

The aim in this chapter is to assist the user by giving these matters an airing. We begin by drawing together broad guidelines relating to power and speed ranges for the various types of motor, then move on to the questions which need to be asked about the load, and finally look briefly at the matter of standards. The whole business of selection is so broad that it really warrants a book to itself, but the cursory treatment here should at least make it easier for the user to arrive at a short-list of possibilities.

POWER RANGE FOR MOTORS AND DRIVES

The diagram (Figure 10.1) gives a broad indication of the power range for the most common types of motor and drive. There is of course no sharp cut-off point in most cases, and this is reflected in the use of dotted lines at the top and bottom of each range. We should also bear in mind that we are talking here about the continuously rated maximum power at the normal base speed, and as we have seen most motors will be able to exceed this for short periods.

Maximum speed and speed range

We saw in Chapter 1 that as a general rule, for a given power the higher the base speed the smaller the motor. There are only a few applications where motors with base speeds below a few hundred rev/min are attractive, and it is usually best to obtain low speeds by means of the appropriate mechanical speed reduction.

Speeds over 10 000 rev/min are also unusual except in small universal motors and special-purpose inverter-fed induction motors. The majority of medium size motors have base speeds between 1 500 and 3 000 rev/min. Base speeds in this range are attractive as far as motor design is concerned since good power/weight ratios are obtained, and are also satisfactory as far as any mechanical transmission is concerned.

In controlled-speed applications, the range over which the steady-state speed must be controlled, and the accuracy of the speed-holding, are significant factors in the selection process.

For constant torque loads which require operation at all speeds the d.c. drive, the inverter-fed induction motor and any of the self-synchronous drives are possibilities, but the latter two may not be capable of continuous operation at low speeds unless the motor is provided with forced cooling.

Fan type loads (see below) with a wide operating speed range are a somewhat easier proposition because the power is

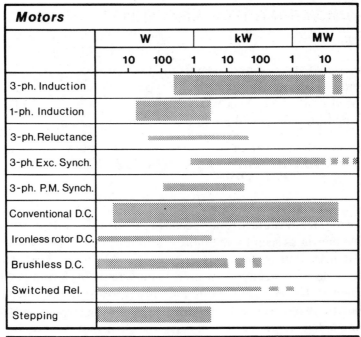

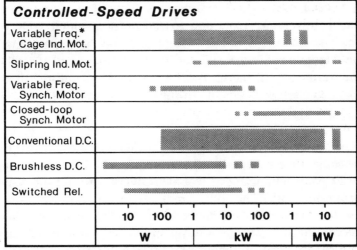

Figure 10.1   *Power ranges for the most common types of motor and drive. *In the low and medium power ranges most of these drives use variable frequency inverters of the type discussed in Chapters 2 and 7, but in the higher power ratings (of a few hundred kW or more) and especially where a restricted speed range is acceptable, a cycloconverter is used to provide the variable-frequency supply*

low at low speeds. In the medium and low power ranges the inverter-fed induction motor (using a standard motor) is satisfactory, and may be cheaper than the d.c. drive. For restricted speed ranges (say from base speed down to 75 per cent) and particularly with fan-type loads where precision speed control is unnecessary, the simple voltage-controlled induction motor is likely to be the cheapest solution.

LOAD REQUIREMENTS – TORQUE-SPEED CHARACTERISTICS

The most important things we need to know about the load are the steady-state torque-speed characteristics, and the effective inertia as seen by the motor. In addition we clearly need to know what performance is required. At one extreme, for example in a steel rolling mill, it may be necessary for the speed to be set at any value over a wide range, and for the mill to react very quickly when a new target speed is demanded. Having reached the set speed, it may be essential that it is held very precisely even when subjected to sudden load changes. At the other extreme, for example a large ventilating fan, the range of set speed may be quite limited (perhaps from 80 per cent to 100 per cent); it may not be important to hold the set speed very precisely; and the times taken to change speeds, or to run-up from rest, are unlikely to be critical.

At full speed both of these examples may demand the same power, and at first sight might therefore be satisfied by the same drive system. But the ventilating fan is obviously an easier proposition, and it would be overkill to use the same system for both. The rolling mill would probably call for a regenerative d.c. drive with tacho or encoder feedback, while the fan could quite happily manage with a cheaper open-loop inverter-fed induction motor drive, or even perhaps a simple voltage-controlled induction motor.

Although loads can vary enormously, it is customary to classify them into two major categories, referred to as 'constant-torque' or 'fan or pump' types. We will use the example of a constant-torque load to illustrate in detail what needs to be done to arrive at a specification for the torque-speed curve. An extensive treatment is warranted because this is often the stage at which users come unstuck.

Constant-torque load

A constant-torque load implies that the torque required to keep the load running is the same at all speeds. A good example is a drum-type hoist, where the torque required varies with the load on the hook, but not with the speed of hoisting. An example is shown in Figure 10.2.

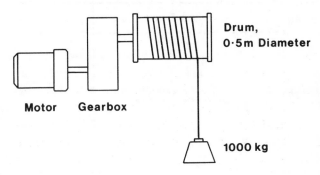

Figure 10.2   *Motor driven hoist*

The drum diameter is 0.5m, so if the maximum load (including the cable) is say 1 000 kg, the tension in the cable will be 9 810 N, and the torque required at the drum for hoisting the maximum load will be 9 810 N × 0.25 m, or approximately 2 500 Nm.

Suppose that the hoisting speed is to be controllable at any value up to a maximum of 0.5 m/s, and that we want this to correspond with a maximum motor speed of around 1 500 rev/min, which is a reasonable speed for a wide range of motors. A hoisting speed of 0.5 m/s corresponds to a drum speed of 19 rev/min, so a suitable gear ratio would be say 80:1, giving a motor speed of 1 520 rev/min.

The load torque, as seen at the motor side of the gearbox,

will be reduced by a factor of 80, from 2500 Nm to 31 Nm at the motor. We must allow also for friction in the gearbox, equivalent to perhaps 20 per cent of the full load torque, so the maximum motor torque required for hoisting will be 37 Nm, and this torque must be available at all speeds up to the maximum of 1520 rev/min.

We can now draw the steady-state torque-speed curve of the load as seen by the motor, as shown in Figure 10.3.

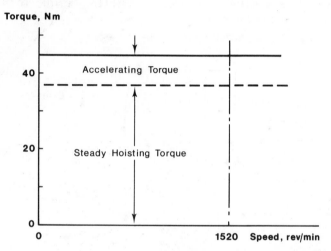

Figure 10.3   *Torque requirements for motor in hoist application (Figure 10.2)*

The motor power is obtained from the product of torque (Nm) and angular velocity (rad/sec). The maximum continuous motor power is therefore given by

$$P_{max} = 37 \times 1520 \times \frac{2\pi}{60} = 5.9 \,\text{kW}. \qquad (10.1)$$

At this stage it is always a good idea to check that we would obtain roughly the same answer for the power by considering the work done per second at the load. The force (F) on the load is 9810 N, the velocity (v) is 0.5 m/s so the power (Fv) is 4.9 kW. This is 20 per cent less than we obtained above, because here we have ignored the power lost in the gearbox.

So far we have established that we need a motor capable of continuously delivering 5.9 kW at 1520 rev/min in order to lift the heaviest load at the maximum required speed. However we have not yet addressed the question of how the load is accelerated from rest and brought up to the maximum speed. During the acceleration phase the motor must produce a torque greater than the load torque, or else the load will descend as soon as the brake is lifted. The greater the difference between the motor torque and the load torque, the higher the acceleration. Suppose we want the heaviest load to reach full speed from rest in say 1 second, and suppose we decide that the acceleration is to be constant. We can calculate the required accelerating torque from the equation of motion, i.e.

$$\text{Torque (Nm)} = \text{Inertia (kgm}^2) \times \text{Ang. Acceln. (rad/sec}^2).$$

$$(10.2)$$

We usually find it best to work in terms of the variables as seen by the motor, and therefore we first need to find the effective total inertia as seen at the motor shaft, then calculate the motor acceleration, and finally use equation 10.2 to obtain the torque.

The effective inertia consists of the inertia of the motor itself, the referred inertia of the drum and gearbox, and the referred inertia of the load on the hook. The term 'referred inertia' means the apparent inertia, viewed from the motor side of the gearbox. If the gearbox has a ratio of n:1 (where n is greater than 1), an inertia of J on the low-speed side appears to be an inertia of J/n^2 at the high-speed side.

In this example the load actually moves in a straight line, so we need to ask what the effective inertia of the load is, as 'seen' at the drum. The geometry here is simple, and it is not difficult to see that as far as the inertia seen by the drum is concerned the load appears to be fixed to the surface of the drum. The load inertia at the drum is then obtained by using the formula for the inertia of a mass m located at radius r, i.e. $J = mr^2$, yielding the effective load inertia at the drum as $1\,000\ kg \times 0.25\,m^2 = 62.5\ kg\ m^2$. Thus the effective inertia of the load as seen by the motor is $1/6\,400 \times 62.5 \approx 0.01\ kg\ m^2$. To this must be added firstly the motor inertia (which we can only estimate by consulting the manufacturer's catalogue for a 5.9 kW, 1520 rev/min motor, which yields a figure of $0.02\ kg\ m^2$), and secondly the referred inertia of the drum and gearbox, which again we have to look up. Suppose this yields a further $0.02\ kg\ m^2$. The total effective inertia is thus $0.05\ kg\ m^2$, of which 40 per cent is due to the motor itself.

The acceleration is easy to obtain, since we know the motor speed is required to rise from zero to 1 520 rev/min in 1 second. The angular acceleration is therefore

$$1\,520 \times \frac{2\pi}{60} \div 1 = 160\,\text{rad/sec}^2$$

We can now calculate the accelerating torque from equation 10.2 as

$$T = 0.05\,\text{kg}\,\text{m}^2 \times 160\,\text{rad/sec} = 8\,\text{Nm}.$$

This means that in order to meet both the steady-state and dynamic torque requirements, a drive capable of delivering a torque of 45 Nm (= 37 + 8) at all speeds is required, as indicated in Figure 10.3.

We note that in this example the torque is dominated by the steady-state requirement, and the inertia-dependent accelerating torque is comparatively modest. Of course if we had specified that the load was to be accelerated in one tenth of a second rather than 1 second, we would require an accelerating torque of 80 Nm rather than 8 Nm, and as far as torque requirements are concerned the system would become dominated by the inertia-dependent accelerating torque, rather than the steady-state running torque.

Fan and pump loads

Fans and pumps have steady-state torque-speed characteristics which generally have the shapes shown in Figure 10.4.

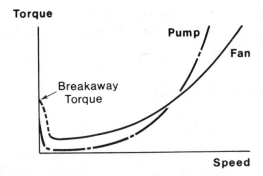

Figure 10.4   *Typical fan and pump-type load characteristics*

These characteristics are often approximately represented by assuming that the torque required is proportional to the square or the cube of the speed, giving rise to the terms 'square-law' or 'cube-law' load. We should note however that the approximation is seldom valid at low speeds because most real fans or pumps have a significant static friction or breakaway torque (as shown in Figure 10.4) which must be overcome when starting.

When we look at the power-speed curves the striking difference between the constant-torque and fan-type load is underlined. If the motor is rated for continuous operation at the full speed, it will be very lightly loaded (typically around 20 per cent) at half-speed, whereas with the constant torque load the power rating will be 50 per cent at half speed. Fan-type loads which require speed control can therefore be handled by drives which can only allow reduced power at low speeds, such as the inverter-fed cage induction motor or the voltage-controlled cage motor. If we assume that the rate of acceleration required is modest, the motor will require a torque-speed characteristic which is just a little greater than the load torque at all speeds. This defines the operating region in the torque-speed plane, from which the drive can be selected.

Many fans do not require speed control of course, and are well served by mains-frequency induction motors. We looked at a typical example in Chapter 6, the run-up behaviour being contrasted with that of a constant-torque load in Figure 6.4.

GENERAL APPLICATION CONSIDERATIONS

Regenerative operation and braking

All motors are inherently capable of regenerative operation, but in drives the basic power converter as used for the 'bottom of the range' version will not normally be capable of continuous regenerative operation. The cost of providing for fully regenerative operation is usually considerable, and users should always ask the question 'do I really need it?'

In most cases it is not the recovery of energy for its own sake which is of prime concern, but rather the need to achieve a specified dynamic performance. Where rapid reversal is called for, for example, kinetic energy has to be removed quickly, and this implies that the energy is either returned to the supply (regenerative operation) or dissipated (usually in a braking resistor). An important point to bear in mind is that a non-regenerative drive will have an asymmetrical transient speed response, so that when a higher speed is demanded, the extra kinetic energy can be provided quickly, but if a lower speed is demanded, the drive can do no better than reduce the torque to zero and allow the speed to coast down.

Duty cycle and rating

This is a complex matter, which in essence reflects the fact that whereas all motors are governed by a thermal (temperature rise) limitation, there are different patterns of operation which can lead to the same ultimate temperature rise. Broadly speaking the procedure is to choose the motor on the basis of the root mean square of the power cycle, on the assumption that the losses (and therefore the temperature rise) vary with the square of the load. For example an induction motor which operates cyclically at full-load (say 4 kW) for two minutes, followed by one minute running light, would have a mean square power of 16/3 and an r.m.s. power rating of 2.3 kW.

Motor suppliers are accustomed to recommending the best type of motor for a given pattern of operation, and they will typically classify the duty type in one of eight standard categories which cover the most commonly encountered modes of operation. As far as rating is concerned the most common classifications are maximum continuous rating, where the motor is capable of operating for an unlimited period, and short time rating, where the motor can only be operated for a limited time (typically 10, 30 or 60 minutes) starting from ambient temperature.

Enclosures and cooling

There is clearly a world of difference between the harsh environment faced by a winch motor on the deck of an ocean-going ship and the comparative comfort enjoyed by a motor driving the drum of an office photocopier. The former must be protected against the ingress of rain and sea water, while the latter can rely on a dry and dust-free atmosphere.

Classifying the extremely diverse range of environments poses a potential problem, but fortunately this is one area where international standards have been agreeed and are widely used. The International Electrotechnical Committee (IEC) standards for motor enclosures are now almost universal and take the form of a classification number prefixed by the letters IP, and followed by two digits. The first digit indicates the protection level against ingress of solid particles ranging from 1 (solid bodies greater than 50 mm diameter) to 5 (dust), while the second relates to the level of protection against ingress of water ranging from 1 (dripping water) through 5 (jets of water) to 8 (submersible). A zero in either the first or second digit indicates no protection.

Methods of motor cooling have also been classified and the more common arrangements are indicated by the letters IC followed by two digits, the first of which indicates the

Plate 10.1   *Flameproof cage induction motor. Motors for use in areas where flammable gases may be present must be designed so that if any gas enters the motor and ignites, the flame is contained inside the motor (Photograph by courtesy of Brook Crompton Parkinson Motors)*

cooling arrangement (e.g. 4 indicates cooling through the surface of the frame of the motor) while the second shows how the cooling circuit power is provided (e.g. 1 indicates motor driven fan).

Dimensional standards

This is another area where standardization is improving, though it remains far from universal. Such matters as shaft diameter, centre height, mounting arrangements, terminal box position, and overall dimensions are fairly closely defined for the mainstream motors (induction, d.c.) over a wide size range, but standardization is relatively poor at the low-power end because so many motors are tailor-made for specific applications.

Supply interaction and harmonics

Converter-fed drives cause distortion of the mains voltage which can therefore upset other sensitive equipment, particularly in the immediate vicinity of the installation. With more and larger drives being installed the problem of mains distortion is increasing all the time, and supply authorities therefore react by imposing increasingly stringent statutory limits governing what is allowable.

There are no agreed international standards at present, and each individual authority sets its own limits. The usual pattern is to specify the maximum amplitude and spectrum of the harmonic currents at various levels in the power system. If the proposed installation exceeds these limits, appropriate filter circuits must be connected in parallel with the installation. These can be costly, and their design is far from simple because the electrical characteristics of the supply system need to be known in advance in order to avoid unwanted resonance phenomena. Users need to be alert to the potential problem, and to ensure that the supplier takes responsibility for handling it.

FURTHER READING

Arcarnley, P. P. (1984) *Stepping Motors: A Guide to Modern Theory and Practice*. Hitchin: Peter Peregrinus.
A comprehensive treatment at a level which will suit both students and users.

Bose, B. K. (1987) *Power Electronics and a.c. Drives*. New Jersey: Prentice Hall.
An in-depth treatment at an advanced level which concentrates on induction and synchronous motor drives, and has two chapters dealing with the use of micro-processors for inverter control.

Hindmarsh, J. (1985) *Electrical Machines and their Applications*. (4th ed.). Oxford: Pergamon.
Hindmarsh, J. (1984) *Electrical Machines and Drives*. (2nd ed.) Oxford: Pergamon.
These two texts by Hindmarsh are popular with both students and practising engineers. The first covers transformers and generators as well as motors, while the second has many worked examples.

Kloss, A. (1984) *A Basic Guide to Power Electronics*. London: Wiley.
Actually an authoritative treatment of line-commutated converters, rather than an introduction to the whole subject of power electronics. It covers all aspects in detail, and will appeal to those who already have some experience.

Murphy, J. M. D. and Turnbull F. G. (1988) *Power Electronic Control of a.c. Motors*. Oxford: Pergamon Press.
A comprehensive treatment of all types of a.c. drive at degree and postgraduate level.

Say, M. G. and Taylor, E. O. (1980) *Direct Current Machines*. London: Pitman.
An established text which covers all aspects of d.c. motors and is written to appeal to both students and practising engineers. Revised and updated to include some material on electronic control.

Sen, P. C. (1981) *Thyristor d.c. drives*. London: Wiley.
A thorough treatment of all types of thyristor d.c. drives at an advanced level. For those who already have some knowledge of d.c. drives.

Staff of Electro-Craft Corp. (1977) *D.c. Motors, Speed Controls, Servo Systems*.
Reprinted and revised many times this book is packed with examples and formulae useful to practising engineers in the servo field.

Thorborg, K. (1987) *Power Electronics*. New Jersey: Prentice Hall.
Wide-ranging and covering the nitty-gritty aspects often overlooked, this book reflects the author's wide industrial experience. Inevitably rather hard going for the absolute beginner.

INDEX